HOW TO THINK ABOUT ANALYSIS

HOW TO THINK ABOUT ANALYSIS

LARA ALCOCK

Mathematics Education Centre, Loughborough University

OXFORD
UNIVERSITY PRESS

OXFORD
UNIVERSITY PRESS

Great Clarendon Street, Oxford, OX2 6DP,
United Kingdom

Oxford University Press is a department of the University of Oxford.
It furthers the University's objective of excellence in research, scholarship,
and education by publishing worldwide. Oxford is a registered trade mark of
Oxford University Press in the UK and in certain other countries

First Edition published in 2014

Reprinted with corrections 2016

Published in the United States of America by Oxford University Press
198 Madison Avenue, New York, NY 10016, United States of America

British Library Cataloguing in Publication Data

Data available

Library of Congress Control Number: 2014935451

ISBN 978-0-19-872353-0

Printed and bound in Great Britain by Clays Ltd, Elcograf S.p.A.

Links to third party websites are provided by Oxford in good faith and
for information only. Oxford disclaims any responsibility for the materials
contained in any third party website referenced in this work.

PREFACE

This preface is written primarily for mathematicians, but student readers might find it interesting too. It describes differences between this book and other Analysis[1] texts and explains the reasons for those differences.

This book is not like other Analysis books. It is not a textbook containing standard content. Rather, it is designed to be read before arriving at university and/or before starting an Analysis course. I really mean that it is designed *to be read*; not to be read like a novel, but to be read at a fair speed. I think this is important in a book that aims to help students make the transition to independent undergraduate study. Students are often unaccustomed to learning mathematics by reading, and research shows that many do not read effectively. This book encourages thoughtful and determined reading without dropping students into so dense a thicket of new definitions and arguments that they become stuck and discouraged.

The book does not, however, fight shy of formality. It contains serious discussions of the central concepts in Analysis, but these begin where the student is likely to be. They examine the student's existing understanding, point out areas in which that understanding is likely to be limited, refute common misconceptions, and explain how formal definitions and theorems capture intuitive ideas in a mathematically sophisticated way. The narrative thus unfolds in what I hope is a natural and engaging style, while developing the rigour of thought appropriate for undergraduate study.

Because of these aims, the book is structured differently from other texts. Part 1 contains four chapters that are about not the content of

---

[1] 'Analysis' should probably not have an upper-case 'A', but I think the multiple everyday meanings of the word mean that it doesn't stand out as a subject name without it.

Analysis but its structure—about what it means to have a coherent mathematical theory and what it takes to understand one. These chapters introduce some notation, but there is no 'preliminaries' chapter. Instead, I introduce notation and definitions where they are first needed, meaning that they are spread across the text (though a short symbol list is provided before the main text, on page xiii). This means that a person reading for review might need to make more than usual use of the index, but I believe this is a price worth paying to give the new reader a smooth introduction to the subject.

A final difference is that not all content is covered at the same depth. The six main chapters in Part 2 contain extensive treatment of the central definition(s), especially where these are logically challenging and where students are known to struggle. They include detailed discussion of selected theorems and proofs, some of which are used to highlight strategies and skills that might be useful elsewhere in a course, and some of which are used to draw out and explain counterintuitive results. Finally, they introduce further related theorems; these are not discussed in detail, but readers are reminded of productive ways to think about them and given a sense of how they fit together to form a coherent theory.

Overall, this book focuses on how a student might make sense of Analysis as it is presented in lectures and in other books—on strategies for understanding definitions, theorems and proofs, rather than for solving problems or constructing proofs. I realize that by taking this approach I risk offending mathematicians, because many value independent construction of ideas and arguments above all else. But three things are clear to me. First, many students scrape through an Analysis course by memorizing large chunks of text with only minimal understanding. This is a terrible situation for numerous reasons, among them that some of those students will go on to be schoolteachers. No one wants to live in a world where mathematics teachers think that advanced mathematics makes no sense—we do not need teachers to reconstruct a subject like Analysis from scratch, but we do want them to understand its main ideas, to appreciate its ingenious arguments, and to inspire their own students to go on to higher study. Second, many students who go on to great things nevertheless suffer an initial period of intense struggle. There is an argument that this is good for them—that, for those who are capable of it, struggling to work things out is better in the long term. I agree with this argument in principle, but I think we should be realistic about its scope. If the challenge is so great that the majority are unable to meaningfully engage, I think we have the balance wrong. Finally, most mathematics lectures are still just that: lectures. Few students follow every detail of a lecture so, no matter the final goal of instruction, an important task for a student is to make sense of written mathematics. Research shows that the typical student is capable of this task to at least some degree, but is ill-informed regarding how to go about it. This book tackles that problem head-on; it aims to deliver students who do not yet know very much Analysis, but who are ready to learn.

A book like this would not be possible without work by numerous researchers in mathematics education and psychology. In particular, the self-explanation training in Chapter 3 was developed in collaboration with Mark Hodds and Matthew Inglis (see Hodds, Alcock & Inglis, 2014) on the basis of earlier research on academic reading by authors including Ainsworth and Burcham (2007), Bielaczyc, Pirolli, and Brown (1995), Chi, de Leeuw, Chiu and LaVancher (1994), and numerous others cited in the bibliography. A pdf version of the training, along with a guide for lecturers, is available free (under a Creative Commons licence) at <http://setmath.lboro.ac.uk>.

Sincere personal thanks go to my friends Heather Cowling, Ant Edwards, Sara Humphries, Matthew Inglis, Ian Jones, Chris Sangwin and David Sirl, all of whom were kind enough to give feedback on earlier versions of various chapters—Chris Sangwin also adapted the Koch snowflake diagrams from <www.ru.j-npcs.org/usoft/WWW/www_tug.org/applications/PSTricks/Fractals/index.html>. Thanks also to the reviewers of the original proposal for their thorough reading and helpful suggestions, and to Keith Mansfield, Clare Charles, Richard Hutchinson, Viki Mortimer and their colleagues at Oxford University Press, whose cheerful and diligent work make the practical aspects of producing a book such a pleasure. Finally, this book is dedicated to David Fowler, who introduced me to Analysis and always gave tutorials with a twinkle in his eye, to Bob Burn, whose book *Numbers and Functions: Steps into Analysis* has greatly influenced my learning and teaching, and to Alan

Robinson, my MSc dissertation supervisor, who told me (at different times) that I should pull my socks up and that I could have a great future writing textbooks. I did pull my socks up, his words stuck with me, and it turns out that writing for undergraduate mathematics students is something I very much enjoy.

CONTENTS

Conclusion 217

SYMBOLS

Symbol	Meaning	Section
$\mathbb{N}$	the set of all natural numbers	Chapter 1
$\forall$	for all	Chapter 1
$\exists$	there exists	Chapter 1
max	maximum	Chapter 1
$\{N_1, N_2\}$	set containing the elements N_1 and N_2	Chapter 1
s.t.	such that	2.2
$\mathbb{R}$	the set of all real numbers	2.2
$\in$	in *or* (which) is an element of	2.2
$f : X \to \mathbb{R}$	function f from X to $\mathbb{R}$	2.4
$\notin$	not in *or* (which) is not an element of	2.5
$\{x \in \mathbb{R} \| x^2 < 3\}$	set of all real x such that $x^2 < 3$	2.6
$[a, b]$	closed interval	2.7
(a, b)	open interval	2.7
$\Rightarrow$	(which) implies (that)	2.10
$\Leftrightarrow$	(which) is equivalent to *or* if and only if	2.10
$\subseteq$	(which) is a subset of	3.2
$X \cup Y$	X union Y	3.2
(a_n)	a general sequence	5.2
ε	epsilon (a Greek letter)	5.5
$\to$	tends to *or* converges to	5.7
$\lim\limits_{n \to \infty} a_n$	limit as n tends to infinity of a_n	5.7
∞	infinity	5.7
Σ	sigma (Greek letter used for a sum)	6.2
$\mathbb{Z}$	the set of all integers	7.3
δ	delta (a Greek letter)	7.4

INTRODUCTION

This short introduction describes the aims and structure of this book, outlines what it covers, and explains how it relates to typical undergraduate Analysis courses.

Analysis is hard. It's elegant, clever, and rewarding to learn, but it's hard. Lots of people will tell you this, including people who are highly successful mathematicians. Ask your lecturers[1] and you will find that a good proportion of them think that Analysis is great now but struggled with it at first. This book will not make it easy—that would be impossible because the logical complexity of the fundamental definitions exceeds that encountered in everyday life and in earlier mathematics, so all Analysis students face an upswing in the demands on their logical reasoning. What this book will do is provide an extended, in-depth explanation of these definitions and of related theorems and proofs. Compared with typical Analysis texts, it gives substantially more attention to the basics, explaining not only the mathematical concepts but also psychological issues associated with learning to think about them. It highlights common errors, misconceptions and sources of confusion—both those that are probably unavoidable due to the subject matter, and those that arise when students over-generalize from their previous mathematical experience. It also explains why some aspects of the formal theory of Analysis can seem odd from a student perspective, but do make sense when you think about them in the right way.

Because of its depth, this book does not provide a lot of content; you will certainly learn more in early Analysis courses than is covered

---

[1] In the UK, where I work, everyone who teaches undergraduate students is referred to as a lecturer. That's not the case everywhere—Americans, for instance, call everyone 'professor'—but it is the language I will stick to in this book.

here. But skills developed by studying the basics in detail can be applied throughout a course, and will provide a solid base from which to tackle more advanced material.

With this in mind, Part 1 focuses explicitly on skills and strategies for learning advanced pure mathematics. This treatment is more condensed than that given in *How to Study for a Mathematics Degree* and its US counterpart *How to Study as a Mathematics Major*, so a student who is new to undergraduate mathematics—or, in a US-style education system, new to upper-level mathematics courses—might want to start with one of those books for broader and more general guidance. This book focuses specifically on Analysis; its illustrations are all drawn from that subject and it contains both detailed information on the structures of mathematical theories and research-based advice on how to study proofs. I recommend that all readers begin with Part 1, even those who already have some experience of undergraduate mathematics—the advice it contains will be referred to throughout.

Part 2 focuses on content, in six areas: sequences, series, continuity, differentiability, integrability and the real numbers. Which areas are most relevant to you will depend upon the Analysis courses at your institution. Some institutions start with sequences and series, then have one or more courses on continuity, differentiability and integrability. Some, however, start with those topics, perhaps in a course that reviews earlier ideas from calculus (roughly, differentiation and integration) and relates them to ideas from Analysis. Work on the real numbers might be included with sequences and series, or might be dealt with elsewhere in a course on foundations or number theory or abstract algebra. Each chapter of Part 2 begins with an overview of its content, so you can compare these with your course specifications and work out which parts to read when.

You might, however, want to read the entire book before studying any Analysis, perhaps in the summer before you begin your undergraduate studies or your upper-level courses. To facilitate this, I have written as though addressing someone who has not yet begun to study the material. However, I hope that the book will also be useful to students who have begun an Analysis course and are struggling to make sense of it, even if they do not start reading until they are preparing for exams.

One important note before beginning: no one should expect to read this entire book fast. Everyone will be able to read *some* of it fast, but the book as a whole is supposed to make you think hard, and any book that does that will, at times, stop you in your tracks. My advice about this would be to read strategically. Have a proper bash at every section but, if you get bogged down somewhere, don't worry about it—just put a sticky-note in the book and move on to the next section, or perhaps to the next chapter. Every chapter contains more and less challenging material, so doing this should get you moving again, and you can always come back to things later.

PART I
Studying Analysis

This part of the book discusses productive ways to think and study when learning Analysis. Chapter 1 is very short—it demonstrates what a page of Analysis notes looks like and gives initial comments on notation and on the form of mathematics at this level. Chapter 2 discusses axioms, definitions and theorems, demonstrating ways to relate abstract statements to examples and diagrams. Chapter 3 discusses proofs—it explains how mathematical theories are structured and provides research-based guidance on how to read and understand logical arguments. Chapter 4 discusses what it feels like to study Analysis, how to keep up, how to avoid wasting time, and how to make good use of resources such as lecture notes, fellow students, and support from lecturers and tutors.

CHAPTER 1

What is Analysis Like?

This chapter demonstrates what definitions, theorems and proofs in Analysis look like. It introduces some notation and explains how symbols and words in Analysis are used and should be read. It points out differences between this type of mathematics and earlier mathematical procedures, and gives initial comments on learning about mathematical theories in a lecture course.

Analysis is different from earlier mathematics, and students who want to understand it therefore need to develop new knowledge and skills. This chapter demonstrates this by showing, on the next page, a typical section of Analysis lecture notes. I do not expect you to understand these notes—the aim of the book is to teach the skills you'll need in order to do that, and Chapter 5 covers the relevant material on sequence convergence. But I do want it to be clear that Analysis is demanding. So turn the page, read what you can, then continue.

Definition: $(a_n) \to a$ if and only if

$$\forall \varepsilon > 0 \; \exists N \in \mathbb{N} \text{ such that } \forall n > N, |a_n - a| < \varepsilon.$$

Theorem: Suppose that $(a_n) \to a$ and $(b_n) \to b$. Then $(a_n b_n) \to ab$.

Proof: Let $(a_n) \to a$ and $(b_n) \to b$.

Let $\varepsilon > 0$ be arbitrary.

Then $\exists N_1 \in \mathbb{N}$ such that $\forall n > N_1, |a_n - a| < \dfrac{\varepsilon}{2|b| + 1}$.

Also (a_n) is bounded because every convergent sequence is bounded.

So $\exists M > 0$ such that $\forall n \in \mathbb{N}, |a_n| \leq M$.

For this M, $\exists N_2 \in \mathbb{N}$ such that $\forall n > N_2, |b_n - b| < \dfrac{\varepsilon}{2M}$.

Let $N = \max\{N_1, N_2\}$.

Then $\forall n > N$,

$$\begin{aligned}
|a_n b_n - ab| &= |a_n b_n - a_n b + a_n b - ab| \\
&\leq |a_n(b_n - b)| + |b(a_n - a)| \\
&\qquad \text{by the triangle inequality} \\
&= |a_n||b_n - b| + |b||a_n - a| \\
&< \frac{M\varepsilon}{2M} + \frac{|b|\varepsilon}{2|b| + 1} \\
&< \frac{\varepsilon}{2} + \frac{\varepsilon}{2} = \varepsilon.
\end{aligned}$$

Hence $(a_n b_n) \to ab$.

Practically every page of your Analysis notes will look like this. On the one hand, that's exciting—you'll be learning some sophisticated mathematics. On the other hand, as you can probably imagine, students who do not know how to interpret such material cannot make sense of Analysis at all. To them, every page looks the same: full of symbols like 'ε', '$\mathbb{N}$', '$\forall$' and '$\exists$', and empty of meaning. By the end of this book, you will be equipped to understand such material: to identify its key components, to recognize how these fit together to form a coherent theory, and to appreciate the intellectual achievements of the mathematicians who created that theory. Right now, I just want to draw your attention to a few important features of the text.

The first feature is that text like this contains a lot of symbols and abbreviations. Here is a list stating what each one means:

(a_n)	a general sequence (usually read as 'a n')
$\rightarrow$	'tends to' or 'converges to'
$\forall$	'for all' or 'for every'
ε	epsilon (a Greek letter, used here as a variable)
$\exists$	'there exists'
$\in$	'in' or '(which) is an element of'
$\mathbb{N}$	the natural numbers (the numbers $1, 2, 3, \ldots$)
max	'(the) maximum (of)'
$\{N_1, N_2\}$	the set containing the numbers N_1 and N_2

Such a list gives you immediate power in that you might not understand the text, but at least now you can read it aloud. Try it: pick a few lines, refer to the list where necessary, and read out what those lines say. You should be able to do this with fairly natural inflections, even if it takes a few attempts. That's because mathematicians write in sentences, so the page might look like a jumble of symbols and words, but it can be read aloud like other text. It might be a while before you can read such material fluently, but fluent reading should be your goal, because if all your energy is taken up with remembering symbol meanings you have little chance of understanding the content. So take opportunities to practise, even if it feels a bit slow and unnatural at first. Don't let lecturers[1] be the only ones who can 'speak' mathematics—aim to own it yourself.

While on the subject of symbols, I would like to comment on their use in this book. Symbols function as abbreviations: they allow us to express mathematical ideas in a condensed form. For this reason, I like them. However, not all lecturers share this view. Some worry that learning new symbols takes up students' mental resources and thereby interferes with their understanding of new concepts. Such lecturers prefer to avoid the symbols and write everything out in words. They are right, of course: it does take a while to get used to new symbols. However, I think that this would be true whenever the symbols were introduced, that there really aren't that many, and that the power they confer makes it worth mastering them early. So I'm going to use them straightaway. I'd like to tell you that I have evidence that this is the best approach, but in this case I don't—it's just personal preference. You can find a full list of symbols used in this book in the Symbols section on page xiii.

The second thing to notice about the page of notes is that it contains a definition, a theorem and a proof. The definition states what it means for a sequence to converge to a limit. This might be far from obvious at this point, but don't worry about that—I'll discuss it in detail in Chapter 5. The theorem is a general statement about what happens when we combine two convergent sequences by multiplying together their respective terms. You can probably see this, and you might be ready to agree that the theorem seems reasonable. The proof is an argument[2] showing that the theorem is true. This argument uses the definition of convergence—notice that some of the symbol strings used in the definition reappear in the proof. The proof starts by assuming that the two sequences (a_n) and (b_n) satisfy the definition, and ends by concluding that the sequence $(a_n b_n)$ satisfies the definition too. It takes some thought to see exactly how the argument fits together, but this book will teach you to look

---

[1] As noted in the Introduction, British people use the word 'lecturer' to refer to anyone who teaches undergraduate students.

[2] When mathematicians say 'argument', they don't mean a verbal fight between two people, they mean a single chain of logically valid reasoning. People use the word in this way in everyday life when they say things like 'That's not a very convincing argument.'

for structures on that level, and I will refer you back to this proof from Section 5.10.

What the page of notes does not contain is a procedure to follow. It is extremely important to recognize this. Students whose mathematical experience to date has consisted mostly of following procedures are often slow to do so. They look for procedures everywhere; they are mystified when they don't find many, and they fail to meaningfully interpret what *is* there. Analysis, like much undergraduate pure mathematics, can be understood as a theory: a network of general results linked together by valid logical arguments known as proofs (see Chapter 3—in particular Section 3.2). The fact that a proof is valid for all objects that satisfy the premises of the associated theorem (see Section 2.2) means that it could be applied repeatedly to particular objects. However, Analysis does not focus on repetitious calculations. Rather, its focus is the theory: it is the theorems, proofs and ways of thinking about them that you are supposed to understand.

The final thing to know is that developing this understanding is your responsibility. You will, of course, have an Analysis lecturer, and maybe an academic tutor or a graduate teaching assistant to offer further face-to-face teaching. These people will do their best to support your learning, but at least some of the time you will be part of a large class where opportunities for individual attention are limited, and you will leave numerous lectures with only a partial understanding of the new material. You therefore need to get good at working out for yourself what it all means. This book is designed to help you do that, starting in the next chapter with some information on the components of mathematical theories.

CHAPTER 2

Axioms, Definitions and Theorems

This chapter is about the building blocks of mathematical theories: axioms, definitions and theorems. It explains their typical logical structures and describes strategies for relating them to examples and diagrams. It illustrates these strategies using Rolle's Theorem and the definition of 'bounded above', and it discusses the utility and limitations of diagrams in general. Finally, it discusses counterexamples and the importance of recognizing the difference between a theorem and its converse.

2.1 Components of mathematics

The main components of a mathematical theory like Analysis are axioms, definitions, theorems and proofs. This chapter discusses the first three of these. Proofs are discussed separately in Chapter 3, but I recommend that you start here, even if you have already begun an Analysis course and you think you are struggling primarily with the proofs—at least some difficulties with proofs arise when people have not fully understood the relevant axioms, definitions and theorems or have not fully understood how a proof should relate to these.

Lots of the axioms, definitions and theorems in Analysis can be represented using diagrams, though people vary in the extent to which they do this. I like diagrams because I find them helpful for understanding abstract information. So I will use a lot of diagrams in this book, and in this chapter I will explain how they can be used to represent both specific and generic examples. I will also offer some words of warning about the limitations of diagrams and the importance of thinking beyond the examples that first come to mind. People who have read *How to Study for/as a Mathematics Degree/Major* will recognize some ideas in this chapter; here the discussion is briefer but more specific to Analysis.

2.2 Axioms

An *axiom* is a statement that mathematicians agree to treat as true; axioms form a basis from which we develop a theory. In Analysis axioms are used to capture intuitive notions about numbers, sequences, functions and so on, so your earlier experience will usually lead you to recognize them as true. They include things like these:

$$\forall a, b \in \mathbb{R}, \ a + b = b + a;$$
$$\exists 0 \in \mathbb{R} \text{ s.t. } \forall a \in \mathbb{R}, \ a + 0 = a = 0 + a.$$

Don't forget to practise reading aloud. Here is a list of the relevant symbols and abbreviations:

$\forall$	'for all' or 'for every'
$\in$	'in' or '(which) is an element of'
$\mathbb{R}$	the real numbers (often read as 'the reals' or simply as 'R')
$\exists$	'there exists'
s.t.	'such that'

Thus, for instance,

$$\forall a, b \in \mathbb{R}, \ a + b = b + a$$

is read as

'For all a, b in the reals, a plus b is equal to b plus a.'

Axioms sometimes have names, so you might see bracketed information before or after each one, like this:

$\forall a, b \in \mathbb{R}, \ a + b = b + a$ [commutativity of addition];

$\exists 0 \in \mathbb{R} \text{ s.t. } \forall a \in \mathbb{R}, \ a + 0 = a = 0 + a$ [existence of an additive identity].

Can you infer the meanings of 'commutativity' and 'additive identity' by looking at these axioms? Can you explain these concepts in your own words without sacrificing accuracy?

Axioms for the real numbers will be discussed in more detail in Chapter 10, which also explains the philosophically interesting shift we make when thinking about mathematical theories in these terms.

2.3 Definitions

A definition is a precise statement of the meaning of a mathematical word. In Analysis you will encounter definitions of new concepts and definitions of concepts that are already familiar. It is the second kind, believe it or not, that will cause you more bother. This is for two reasons. First, some of these definitions will be complicated compared with your existing understanding. They are only as complicated as they need to be and you will come to appreciate their precision, but they take some effort to master and you might have to work through a stage of wondering why things aren't simpler. Second, some of the defined concepts will not quite match your intuitive understanding, so your intuition and the formal theory will occasionally tell you different things, and you will have to sort out the conflict and override your intuitive responses if necessary.

Because of this, I will postpone discussion about definitions of familiar concepts until Part 2. In this chapter I will introduce some definitions of concepts that are likely to be unfamiliar—at least to readers who have not yet studied much undergraduate mathematics—and use these to illustrate skills for interacting with definitions: relating definitions to multiple examples, thinking in terms of diagrams, and being precise.

We will start with the definition below, which I provide in two forms, the first using symbols and the second with (almost) everything written out in words. This should help with your reading aloud, but I'll stop doing it soon so keep up the practice.

Definition: A function $f : X \to \mathbb{R}$ is *bounded above on X* if and only if
$$\exists M \in \mathbb{R} \text{ s.t. } \forall x \in X, f(x) \leq M.$$

Definition (in words): A function f from the set X to the reals is *bounded above on X* if and only if there exists M in the reals such that for all x in X, $f(x)$ is less than or equal to M.

Definitions like this appear routinely in Analysis lectures. They have a predictable structure, and there are two things to notice. First, each definition defines a single concept—this one defines what it means for a certain kind of function to be *bounded above*. In printed material the concept being defined is commonly italicized as here or printed in bold; in handwritten notes, you might see and use underlining instead. Second, this term is said to apply *if and only if* something is true. It is probably easier to see why this is appropriate by considering a simpler definition like this one ('integer' is the proper mathematical name for a whole number):

Definition: A number n is *even* if and only if there exists an integer k such that $n = 2k$.

Splitting this up should enable you to see why both the 'if' and the 'only if' are appropriate:

A number n is even *if* there exists an integer k such that $n = 2k$.
A number n is even *only if* there exists an integer k such that $n = 2k$.

That said, you might see definitions written with just the 'if'. I think this is not ideal, but lots of mathematicians do it because they all know what is intended.

Did you understand the definition of *bounded above*? We will take it apart in detail in the following sections.

2.4 Relating a definition to an example

One way to understand new definitions is to relate them to examples. That sounds simple, but it is important to understand that when mathematicians use the word *example*, they do not usually mean a worked example that shows how to carry out a type of calculation. Rather, they mean a specific object (a function, perhaps, or a number or a set or a sequence) that satisfies a certain property or combination of properties. This can cause miscommunication between lecturers and students. Students say 'We want more examples,' meaning that they want more worked examples, and lecturers think, 'What are you talking about? I've given loads of examples,' meaning examples of objects that satisfy the properties under discussion. Because advanced mathematics is less about learning and applying procedures and more about understanding logical relationships between concepts, worked examples are fewer and further between. And examples of objects are more important—knowledge of a few key examples can clarify logical relationships and help you to remember them. Because of this, your lecturers will almost certainly illustrate definitions using examples. But I want you to develop confidence in generating your own so that you don't have to rely on lecturers for this; in this section and the next I'll describe some ways to go about it.

To get us started, here again is the definition of *bounded above* (if you understood this definition immediately, good, but you might like to read the following explanation anyway as it includes advice on thinking beyond your initial understanding).

Definition: A function $f : X \to \mathbb{R}$ is *bounded above on X* if and only if $\exists\, M \in \mathbb{R}$ s.t. $\forall x \in X, f(x) \leq M$.

This definition defines a property of a function $f : X \to \mathbb{R}$, meaning that f takes elements of the set X as inputs and returns real numbers as outputs. Many people, when asked to think about a function, think about $f(x) = x^2$, so we will start with that. Notice that this function is defined for every real number, so its domain is $X = \mathbb{R}$ and it is a function $f : \mathbb{R} \to \mathbb{R}$. To establish whether or not this function is bounded above, we ask whether or not the definition is satisfied. Substituting in all the appropriate information, $f : \mathbb{R} \to \mathbb{R}$ given by $f(x) = x^2$ is *bounded above on* $\mathbb{R}$ if and only if $\exists\, M \in \mathbb{R}$ s.t. $\forall x \in \mathbb{R}, x^2 \leq M$. Check to make sure you can see this.

So, is the definition satisfied? Does there exist a real number M such that for every real number x, $x^2 \leq M$? Even if you can answer immediately, it is worth noting that it is often easier to start making sense of a definition like this from the end rather than from the beginning. Here the last part says '$f(x) \leq M$', which can be thought of in terms of checking whether values on the vertical axis[1] are less than or equal to M:

---

[1] You might want to call this the y-axis. That's fine, but I will tend to use the notation $f(x)$ instead of y because it generalizes better when working with multiple functions (as we often do in Analysis) or with functions of more than one variable (as we do in multivariable calculus).

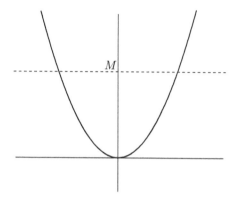

For the M shown, some numbers x in the domain $\mathbb{R}$ have $f(x) \leq M$ and some don't. So for this M it is not true that $\forall x \in \mathbb{R},\ f(x) \leq M$. However, we are interested in whether or not *there exists M* such that $\forall x \in \mathbb{R}, f(x) \leq M$. Does there exist such a number? No. Even for a really big M, there will still be domain values x for which $f(x) > M$. So this function does not satisfy the definition, meaning that it is not bounded above on the set $X = \mathbb{R}$.

2.5 Relating a definition to more examples

To think about a function that *is* bounded above on a set, we can do three things. The first is probably the most obvious: think about different functions. Can you think of a function that is bounded above? Can you think of lots of different ones, in fact? One that might come to mind is $f : \mathbb{R} \to \mathbb{R}$ given by $f(x) = \sin x$, which is bounded above by $M = 1$ because $\forall x \in \mathbb{R}$, $\sin x \leq 1$. It is also bounded above by $M = 2$, notice, because it is also true that $\forall x \in \mathbb{R}$, $\sin x \leq 2$ (there is nothing in the definition to say that M has to be the 'best' bound). We might also consider really simple functions like the constant function $f : \mathbb{R} \to \mathbb{R}$ given by $f(x) = 106$ (meaning that $f(x) = 106\ \forall x \in \mathbb{R}$). This isn't very interesting but it is a perfectly good function, and it is certainly bounded above. Or we might consider $f : \mathbb{R} \to \mathbb{R}$ given by $f(x) = 3 - x^2$. This is bounded above by 3, for instance. It happens not to be bounded below—could you write down a definition of *bounded below* and confirm this? And can you think of a function that is not bounded above and not bounded below?

A second thing we might consider is changing the set X. This is less likely to occur to new undergraduate students, because earlier mathematics tends to involve functions from the reals to the reals. But there is nothing to stop us restricting the domain to, say, the set $X = [0, 10]$ (the set containing the numbers 0, 10 and every number in between). The function $f : [0, 10] \to \mathbb{R}$ given by $f(x) = x^2$ is bounded above on $[0, 10]$ by $M = 100$, because $\forall x \in [0, 10], f(x) \leq 100$. What other numbers could play the role of M here?

Finally, we might stop thinking about specific functions, and instead imagine a generic one. To get a general sense of what this definition says, I might draw or imagine something like this:

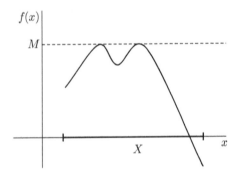

This diagram represents a function on a set X, because for every $x \in X$ (the thickened bar on the x-axis) there is a corresponding $f(x)$. But it isn't supposed to represent a function for which I have a formula in mind, which is appropriate because the definition applies to all functions, not just to those that can be specified by nice formulas. I have also made an effort to capture all aspects of the definition. The diagram shows a restricted set X, for instance, rather than assuming that $X = \mathbb{R}$. In fact, I have drawn a function that is defined only on this set X. Students usually tend to sketch functions that are defined on the whole of $\mathbb{R}$, but that isn't necessary. I have also shown a specific M on the vertical axis, and extended a horizontal line across so that it's clear that all the $f(x)$ values lie below this. Finally, I have made $f(x)$ equal to M in a couple of places to illustrate the fact that this is allowed.

These things all relate to information that appears explicitly in the definition. However, I could add more to the diagram, either to exhibit my own understanding or to explain the definition to someone else. I could, for instance, illustrate the fact that greater values of M are also upper bounds by adding another one, perhaps with some commentary:

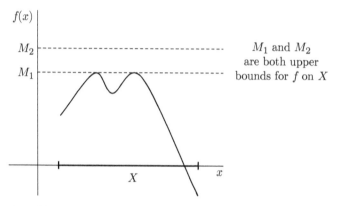

M_1 and M_2 are both upper bounds for f on X

Or I could emphasize the fact that the function is bounded above *on the set* X by extending the graph upwards elsewhere; this would illustrate the idea that the definition says nothing about $f(x)$ values for $x \notin X$ (the symbol '$\notin$' means 'not in'):

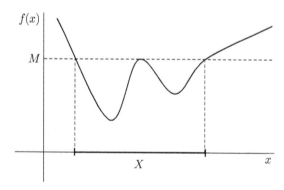

Or I could think about a more complicated set X. You don't have to do this type of exploration, but I think that it provides a fuller sense of the meaning of the definition, which is important because your lecturer will not have time to give an extensive explanation for every concept. Most likely, he or she will introduce a definition and show how it applies (or doesn't) to just one or two examples, assuming that you will do thinking like this for yourself.

2.6 Precision in using definitions

Later chapters will contain information about specific definitions and guidance on how to recognize where these are used in proofs. Here I want to emphasize the importance of precision when working with definitions. To demonstrate what I mean, here is another definition:

Definition: M is an *upper bound* for the function f on the set X if and only if $\forall x \in X, f(x) \leq M$.

This definition and the previous one are about the same core ideas. But the previous one defines what it means for a function to be bounded above on a set—it is about the *function*. This one defines what it means for a number to be a bound for a function on a set—it is about the *number*. This is a subtle distinction but it is one that mathematicians take care over. Imagine that an exam or test asks what it means for M to be an upper bound for a function f on a set X. This requires the second definition, and there are two ways in which a student might fail to answer well. The first is to give an informal answer, saying something like 'It means the function is below M'. When I read this kind of thing I sigh, because the student has clearly understood something about the concept but failed to grasp the fact that mathematicians work with precise definitions.[2] The second is to give the first definition, defining what it means for a function to be bounded rather than what it means for M to be a bound. This would be better but it would still not merit full marks because it doesn't answer the question as asked.

To illustrate how things can go more badly wrong, consider this third definition, which is also to do with boundedness:

[2] For more on why, see Chapter 3 in *How to Study for/as a Mathematics Degree/Major*.

Definition: The set X is *bounded above* if and only if $\exists \, M \in \mathbb{R}$ s.t.
$$\forall x \in X, x \leq M.$$

Students often confuse this with the definition of a function f being bounded on a set X. But look carefully: in this case *there is no function*. This definition is not about a function being bounded above on a set, it is about a *set* being bounded above; it is x-values that are related to M. Here is an example of a set that is bounded above:

$\{x \in \mathbb{R} \mid x^2 < 3\}$ ('the set of all x in R such that x^2 is less than 3').

This set is bounded above by, for example, $\sqrt{3}$, or by 522.

Here is an appropriate generic diagram:

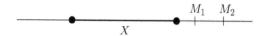

Notice that this definition is just about sets of real numbers so there is no need for a two-dimensional graph—everything of interest can be represented on a single number line. I hope this convinces you that attention to detail is important if we are to distinguish related concepts. And note that the need for precision makes it risky to memorize definitions by rote— much better to understand them properly so you can reconstruct them meaningfully.

2.7 Theorems

A theorem is a statement about a relationship between concepts. Usually this is a relationship that holds *in general*, where I use this phrase in the mathematical sense: when mathematicians say 'in general', they often mean in all cases, not just in the majority of cases.[3] In this section and the next I will explain how to understand theorems by identifying their *premises* and *conclusions* and by systematically seeking examples

---

[3] As a student you should pay attention to differences between everyday and mathematical English so that you do not get confused or misinterpret what someone is saying. If you do, I predict that you will find these differences strange for a couple of months, then you will stop noticing them, then you will become someone who naturally uses words in a mathematical way.

that demonstrate why each premise is needed. We will work with this theorem, which is about functions (the notation is explained below):

Rolle's Theorem:
Suppose that $f : [a, b] \to \mathbb{R}$ is continuous on $[a, b]$ and differentiable on (a, b), and that $f(a) = f(b)$. Then $\exists\, c \in (a, b)$ such that $f'(c) = 0$.

All theorems have one or more *premises*—things that we assume—and a *conclusion*—something that is definitely true if the premises are true. In this case, the premises are flagged by the word 'Suppose'. They are:

- that f is a function defined on an interval $[a, b]$;
- that f is continuous on the interval $[a, b]$;
- that f is differentiable on the interval (a, b);
- that $f(a) = f(b)$.

That's quite a few premises; each will be discussed in detail below.

The conclusion is flagged by the word 'Then'; in this case it is that there exists $c \in (a, b)$ such that $f'(c) = 0$. The notation $f'(c) = 0$ means that the derivative of f at c is zero,[4] and the theorem tells us that there exists a point c in (a, b) with this property (the *open interval* (a, b) is the set containing all the numbers between a and b but not including a or b). The theorem does not tell us exactly where c is—*existence theorems* like this are quite common in advanced mathematics.

As with definitions, we can think about how theorems relate to examples. In this case, we can ask how the theorem applies (or doesn't) to specific functions. To satisfy the premises, a function needs to be defined on a *closed interval* $[a, b]$ (the notation $[a, b]$ means the set containing a and b and every number in between). So we need to decide on a function and on values for a and b as well. For instance, if we take $f(x) = x^2$ with $a = -3$ and $b = 3$, then $f(a) = f(b)$ and f is continuous and differentiable everywhere, so all the premises are satisfied. Thus the conclusion holds: there exists c in (a, b) such that $f'(c) = 0$. In this case the derivative happens to be 0 at $c = 0$, which is certainly between -3 and 3.

Again, you could think of more examples. But I would suggest that when dealing with theorems like this one, you might as well think

---

[4] Many students are more accustomed to the notation df/dx for derivatives, but the $f'(x)$ notation is briefer and is more common in Analysis.

straightaway about a generic diagram. In this case that is doubly beneficial because it requires more careful thought about the premises. To draw a generic diagram, the obvious thing might be to start by drawing a function, but in fact it's often easier to start with the simpler premises. Here, for instance, we can start with points a and b such that $f(a) = f(b)$:

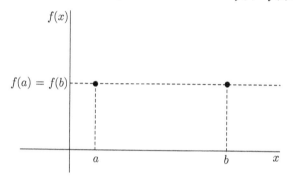

If you just draw what comes naturally in this case you'll end up with a diagram showing a function with the required properties—something like that shown below. Labelling can help to indicate exactly how various parts of the diagram relate to the theorem, so I've labelled an appropriate point c and a little line indicating that the derivative at c is zero. Notice that there are two possible values of c in this diagram, and that it would be straightforward to draw a function that has more.

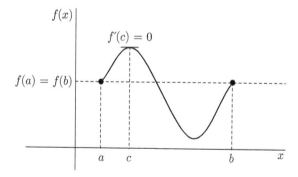

Does the diagram convince you that the theorem is true? Can you see why, given the premises, there must always be a c where $f'(c) = 0$? If your immediate answer is 'yes', that's good, although you might still have something to learn about the technical meanings of continuity and differentiability. If you hesitated because you're not completely sure what we mean by these concepts, that's even better, and you will appreciate the discussion in the next section.

One pernickety point before we move on, though: don't get carried away with your loops when making sketches like this. The diagram below does *not* show a function because there is not a uniquely defined value for $f(d)$, for instance (the graph fails the vertical line test, if you've heard it put that way). I know that when students draw things like this it is usually just carelessness—they usually intend to draw something appropriate. But, again, precision matters in advanced mathematics, so do pay attention to this sort of thing.

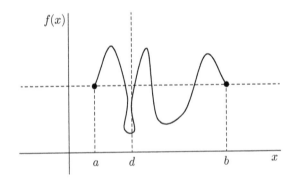

2.8 Examining theorem premises

Rolle's Theorem provides an opportunity to think about the concepts of Analysis in a more serious way, and to learn to think about theorems in depth by asking why all the premises are included. Here is the theorem again:

Rolle's Theorem:
Suppose that $f : [a, b] \to \mathbb{R}$ is continuous on $[a, b]$ and differentiable on (a, b), and that $f(a) = f(b)$. Then $\exists\, c \in (a, b)$ such that $f'(c) = 0$.

One premise is that the function is continuous on the interval $[a, b]$. Most people naturally think about continuous functions, because most functions they have worked with before are continuous (if not everywhere, then at least for most values of x). This book will urge you, however, to avoid thinking only about continuous functions, because the assumption of continuity is sometimes unwarranted—Chapters 7, 8 and 9 include functions that are discontinuous in a variety of interesting ways. Also, we can often gain insight into why a premise is included by thinking about what would go wrong if it were not. Considering Rolle's Theorem, it is quite easy to construct a function that has $f(a) = f(b)$ but that is not continuous on $[a, b]$, and for which the conclusion does not hold. In this diagram, for instance, there is no point c where $f'(c) = 0$:

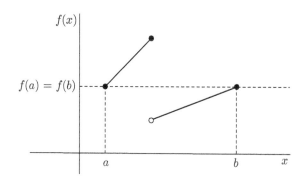

So we need the continuity premise—the theorem would not be true without it. For conceptual insight this diagram is probably enough, but functions like this can be expressed using formulas and it is good practice to give specific examples where possible. Specifying a, b and f like this, for instance, gives a graph like that above:

Let $f : [1, 4] \to \mathbb{R}$ be defined by $f(x) = \begin{cases} x + 1 & \text{if } 1 \le x \le 2 \\ x/2 & \text{if } 2 < x \le 4 \end{cases}$.

This is a *piecewise-defined* function—it is specified differently on different parts of its domain. Notice that it is nevertheless a perfectly good function from $[1, 4]$ to $\mathbb{R}$, because for every x in the interval $[1, 4]$ there is a single specified value of $f(x)$ (sometimes students think this sort of thing is two functions, which is wrong). What do you think is meant by the filled-in dot and the non-filled-in dot in the diagram? Can you come up with other specific examples for which the continuity premise does not hold and the conclusion, again, fails?

Another premise is that the function is differentiable on the interval (a, b). Again, most people naturally think about differentiable functions, because most functions they have worked with before are differentiable. In fact, many new Analysis students are not even aware that they are thinking about differentiable functions, because they have done a lot of differentiation but have not thought in a theoretical way about what it means for a function to be differentiable. Differentiability will be discussed formally in Chapter 8, but for an approximate informal understanding you might think of it as meaning that the graph of the function has no 'sharp corners'. With that in mind, what might go wrong for Rolle's Theorem if the continuity premise holds but the differentiability premise does not? How might the conclusion fail?

Here is a diagram showing one simple case:

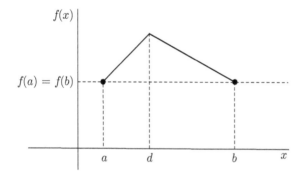

Here $f(a) = f(b)$ and the function is continuous. But it is not differentiable at $x = d$, and nowhere is there a point c with $f'(c) = 0$. I will pause here because, for some readers, it might not be obvious that these things are true. Some students, for instance, are unsure about whether a function like this is continuous at d. They see that 'you can draw it without taking your pen off the page', but they hesitate because they are accustomed to graphs of continuous functions being nice and curvy, not pointy. In fact, this function is continuous, and there will be more about such issues in Chapter 7.

Similarly, some students are unsure about the idea of a derivative at the point d. Again, they are accustomed to thinking about derivatives for functions with nice curvy graphs, and some wonder whether a function like this does have a derivative at the 'corner'. This gets to the essence of differentiability, which is about whether or not we could draw a single sensible tangent line at a point. In this case, we can't (what would be its gradient/slope?[5]), and this issue will be explored in detail in Chapter 8. In the meantime I ask you to take my word for it, and again to note that the differentiability premise is necessary; without it, the conclusion might not hold.

---

[5] People in the UK use the word 'gradient'; people in the US use the word 'slope' to mean the same thing.

One formula specifying a function like that in the diagram is

Let $f : [1,4] \to \mathbb{R}$ be defined by $f(x) = \begin{cases} x+1 & \text{if } 1 \leq x \leq 2 \\ 4 - x/2 & \text{if } 2 < x \leq 4 \end{cases}$.

A simpler example with similar properties would be the function $f(x) = |x|$, on, say, the set $[-5, 5]$. In fact, $f(x) = |x|$ is everyone's favourite example of a function that (at the point $x = 0$) is continuous but not differentiable. You will probably see it introduced as an example of such, but you should bear in mind that when mathematicians give a single, simple example like that, they often intend you to see it as a representative of a general class. They show you $f(x) = |x|$ and perhaps a proof of some claim about it, but they intend that you will generalize the thinking to other functions with similar properties.

2.9 Diagrams and generality

Astute readers might have noticed that I've glossed over three subtleties about diagrams. The first is that, although I've been talking about some diagrams as generic, in a technical sense they are not. As soon as we commit a graph to paper, we are looking at a specific function. However, I think most readers will agree that some diagrams can be thought of as generic in the sense that they are not supposed to call to mind a formula; they don't tempt us to get distracted by knowledge of specific functions in the way that a graph of $f(x) = x^2$ or $g(x) = \sin x$ might.

The second subtlety is that a diagram might not represent a 'whole' function. It is often easy enough to sketch a function on an interval, but a finite diagram cannot fully represent a function defined on the whole of the reals. This probably doesn't bother you, and in most cases it shouldn't: you'll be accustomed to imagining graphs that 'carry on forever' in a predictable way. But do be aware that because any particular diagram is finite (and specific), graphs in and of themselves don't prove very much. They might provide insights that are valuable for constructing proofs, but mathematicians look for definition-based arguments as well.

The third subtlety is that people are often a bit careless about some aspects of their drawings, so they can be misled by local properties of graphs that represent what is going on near $(0, 0)$. For instance, some people tend to sketch a graph of $f : \mathbb{R} \to \mathbb{R}$ given by $f(x) = x^2$ so that it appears to have a U-shape, like this:

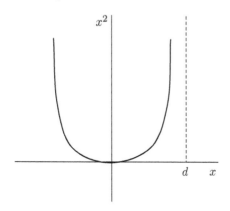

That's misleading because it appears to have vertical asymptotes. What would be the value of x^2 for $x = d$, for instance? It looks like there isn't one, which obviously isn't appropriate. Again, this is a bit pernickety, but you want to draw your graphs so that mathematicians know that you are aware of such issues.

Even when taking care to make the graph look parabolic rather than U-shaped, though, we can be misled by other things. For instance, sketching the graph of f along with that of $g : \mathbb{R} \to \mathbb{R}$ given by $g(x) = x^3$ gives something like this:

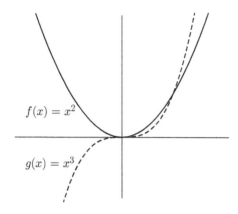

In this diagram, the shapes of the graphs for $x \geq 0$ look more or less alike; it looks like the functions change in a similar way. But, of course, they really don't. We don't have to zoom out much before they start to look very different, as can be seen here:

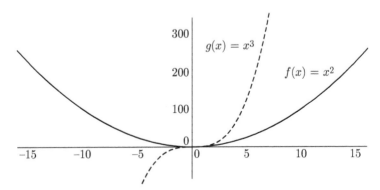

If you tend to be a bit lazy about your sketching, this should make you pause for thought. It should also make everyone think in a more nuanced way about how functions behave for 'big' values.

Similarly, how about comparing the graphs of f and g with that of the exponential function $h : \mathbb{R} \to \mathbb{R}$ given by $h(x) = 2^x$? The function h crosses the vertical axis in a different place but again, other than that, we might be inclined to make them look quite similar:

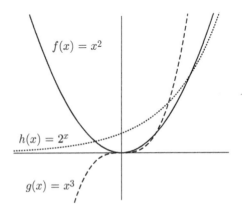

Do you know what happens for bigger values of x, though? The exponential function gets bigger *much* faster. Here is a zoomed-out view that gives a completely different sense of the relative behaviours of these functions:

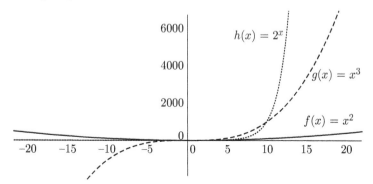

The issues this raises are central in Analysis, where one thing we study is limiting properties: what happens to a function (or a sequence) as x (or n) tends to infinity. Graphs can be useful for developing intuition about this kind of thing, but you should also get out of the habit of treating computer or calculator output as the ultimate arbiter of truth. In Analysis we ask for more than that, developing understanding of the formal theory so that we can actually prove results about our observations.

2.10 Theorems and converses

This final section extends my earlier comments about the structures of theorems and points out some important differences between related *conditional statements*. A conditional statement is an 'if...then...' statement, like this:

If f is a constant function then $f'(x) = 0 \; \forall x \in \mathbb{R}$.

This is a true statement. Here is its *converse*:

If $f'(x) = 0 \; \forall x \in \mathbb{R}$ then f is a constant function.

This is also a true statement. It is not the same statement, though. Here is a *biconditional statement* that captures both things together:

f is a constant function if and only if $f'(x) = 0 \; \forall x \in \mathbb{R}$.

This is also a true statement, but it is a different statement again. You might like to think about the phrase *if and only if* and to consider how this biconditional statement captures both of the conditional statements.

There are two technical things to be aware of at this point. The first is that we could use an alternative notation, writing the three statements like this:

f is a constant function $\Rightarrow f'(x) = 0 \; \forall x \in \mathbb{R}$.
$f'(x) = 0 \; \forall x \in \mathbb{R} \Rightarrow f$ is a constant function.
f is a constant function $\Leftrightarrow f'(x) = 0 \; \forall x \in \mathbb{R}$.

The symbol '$\Rightarrow$' is read aloud as 'implies (that)' and the symbol '$\Leftrightarrow$' is read as 'if and only if' or 'is equivalent to'. These are specific, standard meanings—don't use the arrows unless you intend exactly these meanings.

The second technical thing is that the first conditional statement should really be written like this:

For every function $f : \mathbb{R} \to \mathbb{R}$, if f is a constant function then $f'(x) = 0$ $\forall x \in \mathbb{R}$.

The new bit at the beginning just clarifies that we are talking about all functions of a certain kind. Probably you assumed that in any case, and most mathematicians would too, so extra phrases like this are often omitted. But mathematicians interpret conditional statements as though they were there.

Now, in everyday life we tend to be a bit sloppy in our use of conditional statements. We do not always distinguish a statement from its converse, and we often interpret a conditional statement as though it were a biconditional.[6] In fact, this is so common that there is extensive literature in psychology devoted to people's everyday interpretations of, and reasoning with, conditional statements.

In mathematics, we are not sloppy. When mathematicians write a conditional statement, they mean it exactly as written. This is very important, for two reasons. First, proving a statement is different from proving its converse.

[6] For a detailed explanation see Section 4.6 of *How to Study for/as a Mathematics Degree/Major*.

To prove that

if f is a constant function then $f'(x) = 0 \ \forall x \in \mathbb{R}$,

we would assume that f is a constant function and deduce from this that $f'(x) = 0 \ \forall x \in \mathbb{R}$. To prove that

if $f'(x) = 0 \ \forall x \in \mathbb{R}$ then f is a constant function,

we would assume that $f'(x) = 0 \ \forall x \in \mathbb{R}$ and deduce from this that f is a constant function. This does not necessarily amount to doing the same thing: a proof that works in one direction cannot necessarily be reversed.[7] In this case, one statement can be proved directly from the definition of the derivative, but the other requires more serious theoretical machinery—see Section 8.7 for details.

The second reason is more basic: sometimes a conditional statement is true but its converse is not. For instance, here is another conditional statement:

If f is differentiable at c then f is continuous at c.

This is true. Here is its converse:

If f is continuous at c then f is differentiable at c.

This is **not** true. We have already seen that the function $f(x) = |x|$ is continuous at 0 but not differentiable at 0, meaning that it constitutes a *counterexample* to the conditional statement, demonstrating that it is not universally true. This explains, incidentally, why people have favourite examples of functions and other mathematical objects. Some examples are particularly valuable for remembering key theorems and for avoiding mixing up theorems and their converses. This is handy because Analysis is awash with true theorems that have false converses. Here are a few to be going on with—what is the converse in each case? And do you know enough at present to see why the theorem is true but the converse is not?

---

[7] For a straightforward algebraic example, see Section 8.3 of *How to Study for/as a Mathematics Degree/Major*.

Theorem: If $(a_n) \to \infty$ then $(1/a_n) \to 0$.

Theorem: If $\displaystyle\sum_{n=1}^{\infty} a_n$ converges then $(a_n) \to 0$.

Theorem: If f is continuous on $[a, b]$ then f is bounded on $[a, b]$.

Theorem: If f and g are both differentiable at a then $f + g$ is differentiable at a.

Theorem: If f is bounded and increasing on $[a, b]$ then f is integrable on $[a, b]$.

Theorem: If $x, y \in \mathbb{Q}$ then $xy \in \mathbb{Q}$.

Some of these theorems appear later in the book. Some don't, but you will likely see them in an Analysis course. There might be lots more in your course as well. Whenever you see a conditional statement, I would advise you to think about its converse and think about whether either or both are true; doing so should help you to understand the structure of any accompanying proof. I also have lots more advice about understanding proofs, which you can find in the next chapter.

CHAPTER 3

Proofs

This chapter discusses the meaning of proof in mathematics and the place of proofs in mathematical theories. It discusses ways in which theories and proofs are structured, and ways in which they are taught. It also provides self-explanation training, which has been shown in research studies to improve students' proof comprehension.

3.1 Proofs and mathematical theories

Undergraduate students often think that proofs are mysterious. They're really not. A specific proof might be difficult to understand because of its logical complexity, or because a student doesn't have a good enough grip on the definitions of the relevant concepts. But the idea is not difficult at all: a proof is just a convincing argument that something is true. The apparent mysteries, I think, arise because proofs in a subject like Analysis have to fit within a mathematical theory so, in addition to being convincing, they have to be structured according to the appropriate definitions and theorems. Part 2 of this book is about specific definitions and theorems associated with key concepts in Analysis, about how to identify where they are used in proofs, and about how to use them to structure proofs of your own. In this chapter, I will discuss general strategies for making sense of proofs presented in lectures or textbooks—these strategies can (and should) be applied whenever you encounter a proof in Analysis. First, though, I will briefly explain how proofs fit into mathematical theories.

One thing to get out of the way is that a *theory* is different from a *theorem*. A *theorem*, as discussed in Chapter 2, is a single statement about a relationship between some mathematical concepts. A mathematical *theory* is a network of interconnected axioms, definitions, theorems and proofs. This network might be huge. My current Analysis course contains 16 axioms, 32 definitions and 60 theorems with accompanying proofs. That's not nearly as scary as it sounds, because many of them are very simple. But this is only a small part of what might be considered the 'whole' theory of Analysis. As you might imagine, then, theories can be very complex. But they have some features that make them simpler to understand, and knowing what to look for should help you to appreciate what proofs are for and how they work.

3.2 The structure of a mathematical theory

Mathematical theories are developed over time, and this development is not linear. Mathematicians try to solve problems and to state and prove theorems, and to do so they formulate axioms and definitions to capture the concepts they wish to use. But mathematicians also value theory building—they want everything to fit into a coherent overall structure, which means that axioms, definitions, theorems and proofs are adjusted as groups of mathematicians come to agree on effective ways to capture both individual concepts and important logical relationships.

As a student, you might also solve problems. But unless you take an unusual Analysis course, you will not often be involved in formulating definitions and theorems. Rather, your job will be to learn the established theory of Analysis as it is understood by the contemporary mathematical community. This means that you can think of mathematical definitions and axioms[1] as 'basic' in the sense that they form the bottom layer of the theory's building blocks.

---

[1] See Section 2.2 for a brief introduction to axioms, and Chapter 10 for a more detailed discussion of axioms for the real numbers.

With the bottom layer in place, new blocks at higher levels take the form of theorems, where each theorem says something about a relationship between concepts from the preceding levels. In the initial stages of an Analysis course, theorems might be about just one concept. They will say, for instance, that a property that applies to objects x and y also applies to an object created by combining x and y; by adding them if they are numbers or functions or sequences, for instance, or by taking their union if they are sets. Here are some theorems like that:[2]

Theorem: If $x, y \in \mathbb{Q}$ then $xy \in \mathbb{Q}$.

Theorem: Suppose that $f : \mathbb{R} \to \mathbb{R}$ and $g : \mathbb{R} \to \mathbb{R}$ are both differentiable at a. Then $f + g$ is differentiable at a with $(f + g)'(a) = f'(a) + g'(a)$.

Theorem: If $X, Y \subseteq \mathbb{R}$ are both bounded above, then $X \cup Y$ is bounded above.

Proving such a theorem would involve just one definition. The third theorem, for instance, says that if two subsets X and Y of $\mathbb{R}$ are both bounded above, then their union (the set of all elements in X or in Y or in both) is also bounded above. To prove it, we would do this:

Suppose that X and Y are both bounded above.

Say what this means in terms of the definition of bounded above.

Use algebraic manipulations and logical deductions to construct an argument showing that $X \cup Y$ also satisfies the definition of bounded above.

Conclude that $X \cup Y$ is bounded above.

Because I introduced the definition of *bounded above* in Section 2.6, I will show how the detail is filled in here. Guidance on reading and

---

[2] There is a notation list in the Symbols section at the start of the book, on page xiii.

understanding proofs is provided in Section 3.5; you might like to see what sense you can make of this proof now, then come back to it after that.

Theorem: If $X, Y \subseteq \mathbb{R}$ are both bounded above, then $X \cup Y$ is bounded above.

    Proof: Suppose that X and Y are both bounded above.

        Then $\exists M_1 \in \mathbb{R}$ s.t. $\forall x \in X, x \leq M_1$

        and $\exists M_2 \in \mathbb{R}$ s.t. $\forall y \in Y, y \leq M_2$.

        Now consider $M = \max\{M_1, M_2\}$.

        Then $\forall x \in X, x \leq M_1 \leq M$ and $\forall y \in Y, y \leq M_2 \leq M$.

        So every element of $X \cup Y$ is less than or equal to M.

        So $X \cup Y$ is bounded above.

We could think of a theorem like this as adding a new block to the theory that sits above just one definition, or perhaps that uses one definition and an axiom (maybe an axiom about addition or inequalities).

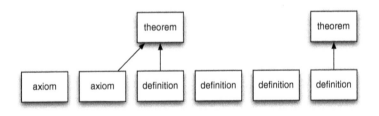

Later theorems will involve multiple concepts. They will say, for instance, that an object with one property must also have another one, or that an object with a combination of properties must also have another one. Here are some theorems like that:

Theorem: Let (a_n) be a convergent sequence. Then (a_n) is bounded.

Theorem: Suppose that $f : [a, b] \to \mathbb{R}$ is continuous on $[a, b]$ and differentiable on (a, b), and that $f(a) = f(b)$. Then $\exists\, c \in (a, b)$ such that $f'(c) = 0$.

Theorem: If f is bounded and increasing on $[a, b]$ then f is integrable on $[a, b]$.

Proving such a theorem would use all the relevant definitions. To prove that every convergent sequence is bounded, for instance, we would do this:

Assume that a sequence (a_n) is convergent.

Say what this means in terms of the definition of convergent.

Use algebraic manipulations and logical deductions to construct an argument showing that (a_n) also satisfies the definition of bounded.

Conclude that (a_n) is bounded.

This time we don't yet have the machinery to fill in the details, but I will return to this theorem in Section 5.9. However, because of its structure, we could think of such a theorem as adding a new block to the theory like this, where again the arrows indicate what was used in the proof:

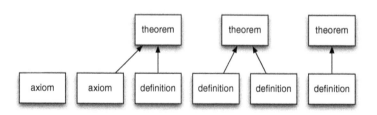

Does this mean that all proofs use definitions directly? No, because once we have proved a theorem, it stays proved. This means that we can use the established theorems to prove new ones, so our theory will build up and up like this:

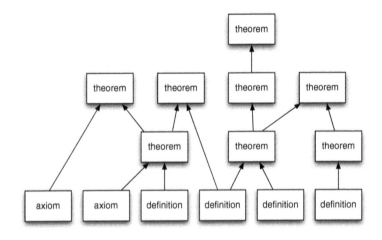

3.3 How Analysis is taught

Although theories are structured as above, it would be a bit weird and disjointed to teach all the definitions and axioms first, then to teach a first 'layer' of theorems, and so on. If we did that, most students would have forgotten the first definition by the time it was used. So lecturers usually introduce a couple of key definitions first and then immediately state and prove some theorems that require only these. They then introduce another definition and build up some more theorems using this together with existing material, and so on. Thinking of theory development in this way should help you to understand the structure of Analysis.

There is, however, one more thing you need to know in order to make sense of an Analysis course. You need to know how this theory fits with earlier mathematics. Most new Analysis students already know a lot of calculus: they know about functions and about how to differentiate and integrate them, and they might also know some things about sequences and series. Many anticipate that Analysis will start from what they know and go upwards; that they will learn fancier and more complicated techniques for integration and differentiation and for working with sequences and series. In fact, that is not what happens at all. Analysis doesn't build upwards from calculus—it sits *underneath* it. In Analysis, we explore the theory that underlies calculus, picking apart our assumptions and understanding why it all works. Other courses do build upwards: mathematics students will probably learn about solving differential equations, and about integration and differentiation for functions of more than one variable or for functions of complex variables. But Analysis, on the whole, builds *down* from calculus, not up.

Analysis doesn't necessarily start with what you know and build down one layer at a time, however. That might make psychological sense, and it might be more reflective of the historical development of the subject, but it's hard to turn it into a logical presentation. Because theories are built on basic axioms and definitions—because proofs use those axioms and definitions—it makes more logical sense to start at the bottom and build back up towards the stuff you already know. So your lecturer might do that and, if so, you're likely to find the experience a bit weird. Analysis will seem a long way from what you've just been studying, and some of the early work will seem insultingly basic. But that's the point: a theory in advanced mathematics should start from basic things and build up a coherent theory.

That said, starting at the very bottom does tend to make students feel disoriented. So, when I teach Analysis, I tend to operate a sort of hybrid in which we start with definitions and study those in detail, but I don't mention axioms at first. Rather, I just get on and use axioms that I think students will take for granted (I'm right—I've never had anyone complain that we haven't specifically stated that $\forall a, b \in \mathbb{R}, a + b = b + a$). Then, once the class has had a crack at proving things on the basis of definitions, we might take another step downward and examine the more basic axiomatic assumptions. Students tend to be more ready for it by then, because they're able to appreciate the importance of systematic reasoning within a network of results. Your lecturer, of course, might do something different.

3.4 Studying proofs

The preceding sections should clarify the roles of definitions, theorems and proofs within a mathematical theory. On a page, however, definitions and theorems are small—typically one line or two—and proofs are bigger—perhaps five lines, or ten or fifteen. As a result, proofs tend to draw the eye and to seem more important, and students often talk about proofs as though they have an independent existence. But a proof is always a proof *of* something, and the something will be stated as a theorem (though it might be called a *proposition* or a *lemma* or a *claim*). Unsurprisingly, then, if you don't understand the theorem, you won't understand its accompanying proof; if you don't know what the author of the proof was trying to establish, how will you know when they've convinced you that they've done it?

So, don't think of a proof as an isolated entity, think of it as belonging to a theorem, and make sure that you understand what the theorem says first. This will often involve thinking at two levels, the first intuitive and the second formal. For instance, if a theorem says that every convergent sequence is bounded, you might find that you have an immediate intuitive sense of what this means. Nevertheless, it is advisable to pause and think properly about what it means in relation to the formal definitions of *convergent* and *bounded*: these are technical concepts, and the theorem is about those technical concepts, not about your overlapping but probably slightly woolly intuitive understanding. See Sections 2.7 and 2.8 for more on how you might do this.

Once you understand what the theorem says, you are in a position to study the proof. But how should you do this? How will you know when something is proved? For many undergraduate students, the obvious answer is that you know something is proved when your lecturer or textbook says it is. Obviously you have no reason to doubt a presented proof: it's entirely reasonable to believe something on the basis that someone in authority tells you that it's valid. It's not very intellectually satisfying, though—much better to understand something in detail than just to believe it. The good news from research in mathematics education is that students generally seem to have enough knowledge and logical reasoning skill to develop pretty good understanding of undergraduate proofs; the bad news is that many of them don't mobilize their knowledge very well. They can, however, do better once they have had some simple *self-explanation training*. Mathematics-specific self-explanation training appears in the next section.

3.5 Self-explanation in mathematics

At my university, we have used self-explanation training in several research studies, with positive results. The training is available at <http://setmath.lboro.ac.uk> and it is reproduced below as used in the studies—I have added two footnotes to link to ideas from elsewhere in the book, but other than that I haven't changed anything except the formatting. Because of this, both the style and the content in this section are a bit different—the style is less conversational and more instructional, and the content is more general—it involves concepts from number theory as well as Analysis.

SELF-EXPLANATION TRAINING

The self-explanation strategy has been found to enhance problem solving and comprehension in learners across a wide variety of academic subjects. It can help you to better understand mathematical proofs: in one recent research study students who had worked through these materials before reading a proof scored 30% higher than a control group on a subsequent proof comprehension test.

HOW TO SELF-EXPLAIN

To improve your understanding of a proof, there is a series of techniques you should apply.

After reading each line:

- Try to identify and elaborate the main ideas in the proof.
- Attempt to explain each line in terms of previous ideas. These may be ideas from the information in the proof, ideas from previous theorems/proofs, or ideas from your own prior knowledge of the topic area.
- Consider any questions that arise if new information contradicts your current understanding.

Before proceeding to the next line of the proof you should ask yourself the following:

- Do I understand the ideas used in that line?
- Do I understand why those ideas have been used?
- How do those ideas link to other ideas in the proof, other theorems, or prior knowledge that I may have?
- Does the self-explanation I have generated help to answer the questions that I am asking?

Below you will find an example showing possible self-explanations generated by students when trying to understand a proof (the labels '(L1)' etc. in the proof indicate line numbers). Please read the example carefully in order to understand how to use this strategy in your own learning.

EXAMPLE SELF-EXPLANATIONS

Theorem: No odd integer can be expressed as the sum of three even integers.

Proof: (L1) Assume, to the contrary, that there is an odd integer x, such that $x = a + b + c$, where a, b, and c are even integers.

(L2) Then $a = 2k, b = 2l$, and $c = 2p$, for some integers k, l, and p.

(L3) Thus $x = a + b + c = 2k + 2l + 2p = 2(k + l + p)$.

(L4) It follows that x is even; a contradiction.

(L5) Thus no odd integer can be expressed as the sum of three even integers.

After reading this proof, one reader made the following self-explanations:

- 'This proof uses the technique of proof by contradiction.'[3]
- 'Since a, b and c are even integers, we have to use the definition of an even integer, which is used in L2.'

---

[3] Proof by contradiction is discussed along with other types of proof in Chapter 6 of *How to Study for/as a Mathematics Degree/Major*.

- 'The proof then replaces a, b and c with their respective definitions in the formula for x.'
- 'The formula for x is then simplified and is shown to satisfy the definition of an even integer also; a contradiction.'
- 'Therefore, no odd integer can be expressed as the sum of three even integers.'

SELF-EXPLANATIONS COMPARED WITH OTHER COMMENTS

You must also be aware that the self-explanation strategy is not the same as *monitoring* or *paraphrasing*. These two methods will not help your learning to the same extent as self-explanation.

Paraphrasing

'a, b and c have to be positive or negative, even whole numbers.'

There is no self-explanation in this statement. No additional information is added or linked. The reader merely uses different words to describe what is already represented in the text by the words 'even integers'. You should avoid using such paraphrasing during your own proof comprehension.[4] Paraphrasing will not improve your understanding of the text as much as self-explanation will.

Monitoring

'OK, I understand that $2(k + l + p)$ is an even integer.'

This statement simply shows the reader's thought process. It is not the same as self-explanation, because the student does not relate the sentence to additional information in the text or to prior knowledge. Please concentrate on self-explanation rather than monitoring.

[4] I don't intend this to contradict the advice in Chapter 1 about reading mathematics aloud. You might need to read what is literally on the page first, but you should then think beyond it to self-explanation.

A possible self-explanation of the same sentence would be:

'OK, $2(k + l + p)$ is an even integer because the sum of 3 integers is an integer and 2 times an integer is an even integer.'

In this example the reader identifies and elaborates the main ideas in the text. They use information that has already been presented to understand the logic of the proof.

This is the approach you should take after reading every line of a proof in order to improve your understanding of the material.

PRACTICE PROOF 1

Now read this short theorem and proof and self-explain each line, either in your head or by making notes on a piece of paper, using the advice from the preceding pages.

Theorem: There is no smallest positive real number.

Proof: Assume, to the contrary, that there exists a smallest positive real number.

Therefore, by assumption, there exists a real number r such that for every positive number s, $0 < r < s$.

Consider $m = r/2$.

Clearly, $0 < m < r$.

This is a contradiction because m is a positive real number that is smaller than r.

Thus there is no smallest positive real number.

PRACTICE PROOF 2

Here's another more complicated proof for practice. This time, a definition is provided too. Remember: use the self-explanation training after *every* line you read, either in your head or by writing on paper.

Definition: An *abundant* number is a positive integer n whose divisors add up to more than $2n$.

For example, 12 is abundant because $1 + 2 + 3 + 4 + 6 + 12 > 24$.

Theorem: The product of two distinct primes is not abundant.

Proof: Let $n = p_1 p_2$, where p_1 and p_2 are distinct primes.

Assume that $2 \leq p_1$ and $3 \leq p_2$.

The divisors of n are $1, p_1, p_2$ and $p_1 p_2$.

Note that $\dfrac{p_1 + 1}{p_1 - 1}$ is a decreasing function of p_1.

So $\max \left\{ \dfrac{p_1 + 1}{p_1 - 1} \right\} = \dfrac{2 + 1}{2 - 1} = 3$.

Hence $\dfrac{p_1 + 1}{p_1 - 1} \leq p_2$.

So $p_1 + 1 \leq p_1 p_2 - p_2$.

So $p_1 + 1 + p_2 \leq p_1 p_2$.

So $1 + p_1 + p_2 + p_1 p_2 \leq 2 p_1 p_2$.

REMEMBER . . .

Using the self-explanation strategy had been shown to substantially improve students' comprehension of mathematical proofs. Try to use it every time you read a proof in lectures, in your notes or in a book.

That's the end of the self-explanation training.[5] Some readers might like to apply it now to the proof in Section 3.2.

3.6 Proofs and proving

This chapter is about studying proofs as they appear in lecture notes or books. Lots of your activity as an undergraduate will involve understanding proofs from such sources. But this does not mean that mathematics is fixed and finished. On the contrary, it is a constantly evolving subject. It happens that Analysis as the mathematical community now understands it was developed (mostly) in the nineteenth century, so it is a fair while since anyone disagreed about its finer details, and modern textbooks all capture the central ideas in essentially the same ways. You will therefore learn about Analysis as a network of results established using standard proofs. But that does not mean that proofs are unique: there might be numerous possible proofs for a single theorem, each using different but valid reasoning. And it does not mean that there is no room for creativity in mathematics. Today, ideas at the current boundaries of mathematics are constructed, compared and debated in thousands of universities around the world. Certainly any student learning in an established area still has plenty of opportunities to solve problems and to develop knowledge independently.

CHAPTER 4

Learning Analysis

This chapter explains what it feels like to study Analysis. It offers advice on how to keep up, how to avoid wasting time, and how to make good use of study resources.

4.1 The Analysis experience

Here is what happens when I teach Analysis. In week 1, everyone is in a good mood because they're starting something new. In weeks 2 and 3, there is a buildup of increasingly challenging material. In week 4, the mood in the lecture theatre is dreadful. The whole class has realized that this is difficult stuff and that it isn't going to get any easier. Everyone hates Analysis and, by extension, quite a few people hate me. I am not fazed by this, though, because I have taught Analysis about twenty times now and I know what will happen next. In week 5, everyone will feel slightly better, even if no one can quite explain why. In week 7, a small number of people will approach me and tell me shyly that, although Analysis is challenging, they're starting to think they might like it. By the end of the course, these people will be telling anyone who will listen that Analysis is brilliant, and lots of other students will admit that now that they're getting the hang of it, they can see why people think it's a great subject.

The question for a new student, then, is how to handle it when the work gets difficult and you start to feel negative. Some students turn the negativity inwards: they lose confidence, experience self-doubt about their mathematical ability ('Perhaps I'm not good enough for this?'), and sometimes become withdrawn. Others turn it outwards, expressing frustration and anger about their lecturers ('He's a terrible teacher!') and sometimes, a bit nonsensically, about the mathematics itself ('I don't know why they're teaching us this rubbish—this isn't maths!'). These reactions both arise naturally when people feel a loss of control and consequently get defensive. But neither is very productive. So what is the alternative?

Well, most people do experience a bit of difficulty when first learning Analysis. This is just a fact of life. So, in my view, the trick is simply to expect this as a normal part of the learning experience, and ride it out. If you are ready for a bit of a challenge, you'll be better placed to handle the emotions without hiding away or acting out—you can say to yourself 'Well, okay, I was expecting this,' and continue to study in a sensible way, knowing that things will gradually come together. This chapter is about practical approaches to doing that.

4.2 Keeping up

In Analysis, as in any undergraduate mathematics course, the big challenge is keeping up. If you're taking a decent course, this will be difficult. No one is trying to teach you stuff that you will find easy—what would be the point of that? Also, you will be busy, with other courses and with the rest of your life. So it is very unlikely that you will be on top of everything all the time. You should try not to be distressed by this, because distress doesn't help—negative emotions just impede effective study. The thing to do is to accept that you will not always have perfect knowledge of everything, and work in an intelligent way that allows you to maintain *sufficient* knowledge of the *important* things.

When I say *sufficient* knowledge, I mean enough knowledge to give you a fighting chance of making sense of new material. By the time you are a few weeks into a course, you are unlikely to understand everything in every lecture—I certainly didn't. But you want to have enough under your belt that you can follow the big sweep of the theory development and understand some of the details. When I say the *important* things, I mean the central concepts that come up again and again. At any given time, it is unlikely that you will be able to explain the nuances of every proof, but you want to know the main definitions and theorems so that you can recognize when and how they are used in new work. With that in mind, here is what I would prioritize.

First, you absolutely must know your definitions. In Analysis, it is sometimes tempting to be lax about this, because many of the words used ('increasing', 'convergent', 'limit' etc.) have everyday meanings, and because concepts in Analysis can often be represented using diagrams. Both of these things will tempt you into thinking that intuitive understanding is sufficient. *It isn't*. Definitions are central to any theory in advanced mathematics: as explained in Chapters 2 and 3, they are key to understanding what is really meant by the theorems and what is going on in many of the proofs. If you think you understand the subject without knowing your definitions properly, you are kidding yourself. Because of this, I would start a definitions list on the first day of the course. Keep this on a piece of paper at the front of your folder (even if you keep most of your notes on an electronic device, I would still use paper for this). Every time you encounter a new definition, add it to the list. Study the list regularly, perhaps test yourself on it periodically, and be alert in lectures for defined words—every time lecturers use one, they mean it in exactly the sense captured by the definition.

Second, it is a good idea to be conversant with the main theorems. These capture relationships between concepts, so knowing what they say—even if you don't fully understand the proofs—will give you an overview of the course. Sections 2.7 and 2.8 give advice on thinking thoroughly about theorem meanings—a few minutes spent following this advice is likely to fix a new theorem in your mind. Also, notes are sometimes provided in advance these days, either for a whole course or for some block of it (if your course follows a textbook, you will have the whole thing in advance). So you could get a sense of what theorems are coming by reading ahead. Once the course gets going, I would consider keeping a theorems list too. Indeed, I would go beyond list-making and construct a *concept map* (sometimes called a *mind map* or a *spider diagram*). Because of the way theory is built up, it often makes sense to use a diagram to indicate which theorems (and definitions) are used to prove which other theorems, as discussed in Section 3.2. You could make a concept map that looks something like this, with the words in the boxes replaced by names or abbreviations for the specific definitions and theorems in your course:

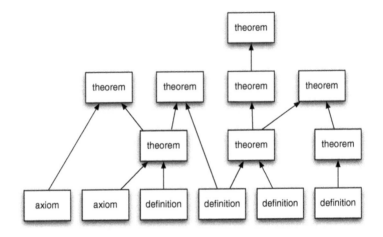

Those are the things I would prioritize. If you find yourself getting behind or otherwise short of time, do those before you do anything else. Don't try to go back to where you last understood everything and work forwards from there—doing so will be ineffective, because the course will move on faster than you do, and you will end up in lectures where you do not understand anything. Analysis is hierarchical and thus unforgiving of students who do not keep up with the main building blocks. Prioritizing as suggested here will usually give you sufficient knowledge of the important things; it will allow you to identify key concepts and relationships in new lectures, and will provide you with a framework for more detailed study, as discussed below.

4.3 Avoiding time-wasting

Because keeping up is challenging, you do not want to waste any time. So it is worth thinking about how much study time you've got and how you're going to use it.[1] In the UK, where I work, lecture courses in subjects like Analysis typically involve about three one-hour lectures per

---

[1] Chapter 11 in *How to Study for/as a Mathematics Degree/Major* discusses large-scale time management for mathematics students; this chapter discusses specific study strategies for an Analysis course.

week. In such a system, I think it reasonable to spend a further three to four hours per week on independent study. If you do that for all your subjects, you'll probably end up with a standard 40-hour working week, which is about right (if you are studying in a different system, you can read the advice below and work out how to adjust it for your situation).

Now, three to four hours isn't very much. You can tell yourself you're going to do more than that if you like, but most people don't, so it's probably more important to make the three to four hours count. During that time, you will have two things to do: study your lecture notes (or your textbook), and work on problems. I put the tasks in that order for a reason. In order to work effectively on the problems, you will need to be familiar with the material in your notes. If you are, you will find that many problems make you think 'Ah, we did something related to this on Wednesday.' If you aren't, you will waste a lot of time having no idea how to start and staring into space. So, notes first.

I suggest spending perhaps sixty to ninety minutes studying your recent notes. This does not mean reading them without really thinking, though. Read everything carefully, following the study suggestions for definitions, theorems and proofs from Chapters 2 and 3, and updating your definitions and theorems lists as you go along. Aim for good self-explanations (see Section 3.5), but don't obsess over anything. Sixty to ninety minutes isn't that long, and you want to be at least somewhat familiar with everything. So, if you have spent a few minutes thinking properly about something but you still don't really get it, get out a piece of paper, write 'Questions about Analysis' at the top, and make a note of where this thing is and what exactly you don't understand. Be precise—sometimes nailing down the problem allows you to sort it out and, if it doesn't, you will have a specific note to come back to so you don't lose the thinking you have already done.

Once you've studied your notes, begin work on the problems. Depending on how you divide up your time, you'll have between two and three hours for this. That will not give you very much time for any given problem, so again you don't want to waste any. Because of this, I suggest a first pass in which you spend perhaps ten minutes on each problem. Some problems you will be able to finish in this time, especially if they involve routine warm-up exercises or direct applications of an idea that you've just studied. (In such cases, see how much you can do without looking at your notes—this might take slightly longer but, if you can construct or reconstruct something for yourself, you will remember it better in the long run.) Other problems you will not be able to finish in ten minutes. If you are making good progress, you might want to carry on for a bit longer. If, on the other hand, you're stuck, and if you've tried a few sensible things to get unstuck, make a note on your 'Questions about Analysis' sheet and move on—those other problems are still waiting.

Now, I said 'at a first pass' because I think problem solving in Analysis should be a multiple-pass task. You want to have a go, then have a break for a day or two, then have another go. Magical things will sometimes happen in the break—your brain will make new connections and you'll see new ways forward. So you probably want to break up your study into at least a couple of blocks. Indeed, you should do that anyway, because thoughtful study is intellectually effortful—if you decide to spend four hours at a stretch studying Analysis, I guarantee that you will waste the last two simply because you will run out of energy.

4.4 Getting your questions answered

Next, what to do with your 'Questions about Analysis' list? For a start, keep an eye on it. Sometimes, working on problems will make you think about an idea in a different way, and you'll be able to cross off something that you added when studying your notes. Sometimes, when you've had a break for a couple of days, a quick re-read of your notes will make something click, and you'll be able to finish a problem and cross that off too. After that, here's what I'd do.

First, get together with a friend or two and work systematically through your respective lists. Everyone thinks a bit differently, so you will probably be able to fill some gaps for one another. Doing this will also force you to speak about Analysis, helping you to become fluent in talking about the concepts and explaining your arguments. Fluency is important, so don't worry if you trip over your words at first. Just have another go—you will only get more confident with practice. Sharing ideas will also help you to become a good mathematical listener. Pay close attention to what your friends are saying and, if you are not sure you understand, say so, and try to specify what is confusing you. Doing this will help your friends to articulate their thoughts more clearly. Again, this is a valuable skill that will help all of you to speak more confidently to lecturers and other tutors. Of course, as in individual work, don't get obsessed—if you can't sort something out between you in a reasonable amount of time, perhaps your effort would be better spent elsewhere.

Once you've shared your knowledge with friends, take your remaining questions to an expert (you can always go to the expert first, of course, but consider the above issues about developing communication skills). Which expert you want will depend on your institution's teaching systems: perhaps your tutor, perhaps your course lecturer, perhaps a mathematics support service. Whoever you see, take your list and your problem sheets and all your relevant notes, and make sure that your list has page or section or question numbers on it—you want to be able to find everything with minimal fuss. If seeing someone involves arranging a specific meeting, consider asking whether you and your friends can go together—that should make the process more efficient. And do not be shy about asking questions, even if you have a long list. Trust me, a student asking specific questions from a well-organized list is always impressive.

Taking this approach should mean that most of your questions are answered most of the time. However, do be realistic. Following this advice will still leave you with gaps. Sometimes there will not be time to sort everything out. Sometimes there will be time to sort everything out, but two weeks later you will realize that you've now forgotten why something works and you need to think it through again. It should be possible to minimize that problem by making decent notes—when you've overcome confusion about something, recording how you changed your thinking will facilitate quick review. Overall, if you get yourself organized at approximately the level suggested in this chapter, you will keep up with the main ideas, you will understand at least some of each new lecture, and you will develop a solid block of knowledge that you can build on when you start preparing for exams.

4.5 Adjusting your strategy

In this chapter I have suggested a specific way to organize your studies. I should say that I don't really expect anyone to behave in precisely this way. You will be subject to constraints about when you can study, to personal preferences about your work habits, and to the shifting requirements of other aspects of your academic work and your social life. So you should reflect occasionally on how things are going, and be ready to adjust. If you need longer to study your notes, adjust your timings; if you need time to study for a test in another subject, cut back to the essentials in Analysis for a week; if one of your friends is great for social outings but a bit rubbish at concentrating on Analysis, quietly make alternative or extra arrangements for discussions with others. And of course, if you're really into a problem, stare into space and think about it for hours, if you like. The advice here should be thought of as a useful place to start, and as a way to develop a routine that will keep you going through the challenging weeks.

PART 2
Concepts in Analysis

This part of the book contains six main conceptual chapters providing detailed introductions to advanced work on sequences, series, continuity, differentiability, integrability and the real numbers. Each of these chapters begins by discussing a typical new Analysis student's existing relevant knowledge; it then reframes this knowledge in a more sophisticated way by introducing key definitions, highlighting and resolving common misconceptions and sources of confusion, relating new ideas to examples and diagrams, and raising questions for the reader. Later parts of these chapters discuss selected theorems and proofs, relating these to broader mathematical principles and indicating how they fit into a typical Analysis course. The concluding chapter provides a short review of important things to remember when studying Analysis.

CHAPTER 5

Sequences

This chapter introduces sequence properties such as monotonicity, bounded-ness and convergence, using diagrams and examples to explain these and demonstrating how their definitions are used in various proofs. It discusses questions that arise for sequences that tend to infinity, and describes ways in which this content would fit into a typical Analysis course.

5.1 What is a sequence?

A sequence is an infinite list of numbers, like this:

$$2, 4, 6, 8, 10, 12, \ldots$$

or like one of these:

$$1, \tfrac{1}{3}, \tfrac{1}{9}, \tfrac{1}{27}, \tfrac{1}{81}, \tfrac{1}{243}, \ldots$$

$$1, 0, 1, 0, 1, 0, 1, 0, \ldots$$

Analysis involves the study of various sequence properties and the relationships between them. To think flexibly about those relationships, it helps to be aware of some different ways of representing sequences and some advantages and disadvantages of those representations. Even with this simple list representation, there are a few things to notice.

First, the list has a comma between each pair of sequence terms and another after the last term that is explicitly listed. This is just notational convention, but it's the kind of thing that looks professional if you get it right.

Second, the list ends with an ellipsis—a set of three dots. This is a proper punctuation mark, and here it means 'and so on forever'. It is important to include the ellipsis—otherwise a mathematically educated reader will assume that the list stops at the last stated term, which is inappropriate because in Analysis the word 'sequence' always refers to an infinite sequence. This is not the case in everyday life, where the word 'sequence' might refer to a finite list. As with all definitions in undergraduate mathematics, you are free to think that you prefer the everyday interpretation, but you will have to adhere to the convention in your studies.[1]

Third, the sequence is infinite 'only in one direction'. For instance, this is not a sequence:

$$\ldots, -6, -4, -2, 0, 2, 4, 6, \ldots$$

Another informal way to say this is that a sequence must have a first term. It might seem odd to remark on this, but some situations tempt students to allow sequences to be infinite 'in both directions'. I will point one out in Section 5.9.

Finally, the sequences above follow obvious patterns, but that is not a necessary feature. An infinite list of randomly generated numbers would be a perfectly good sequence. Of course, it would be difficult to work with, so in practice you will mostly see sequences that follow some sort of pattern. But general theorems about sequences apply to all sequences that satisfy their premises,[2] not just to those that are expressible using nice formulas.

5.2 Representing sequences

The comment above notwithstanding, formulas often are useful for representing sequences. The sequence $2, 4, 6, 8, 10, 12, \ldots$, for instance, might be specified by writing this:

Let (a_n) be the sequence defined by $a_n = 2n \ \forall n \in \mathbb{N}$.

---

[1] See Chapter 2 of this book and, for an extended discussion, Chapter 3 of *How to Study for/as a Mathematics Degree/Major*.

[2] See Sections 2.7 and 2.8 for a discussion of theorem premises.

Think about the link between such a specification and the fact that a sequence must have a first term. The set $\mathbb{N}$ of natural numbers is the set $\{1, 2, 3, 4, \ldots\}$, so this specification yields $a_1 = 2$, $a_2 = 4$, and so on; there is no a_0 or a_{-1}. Note also that a_n denotes the nth term of the sequence and (a_n) denotes the whole sequence. These are very different—a_n is a single number and (a_n) is an infinite list of numbers—so make sure you write the one you intend. An alternative notation for the whole sequence is $\{a_n\}_{n=1}^{\infty}$. I'm not keen on that one, partly because it is longer, and partly because curly braces are also used to denote sets in which the order of the listed terms does not matter. For sequences, the order does matter. So I will stick with the round-bracket version for this book, but you should adopt whichever your course or textbook uses.

To abbreviate further we can write a formula in the brackets, as in sentences like these:

Consider the sequence $(2n)$.

The sequence $\left(\dfrac{1}{3^{n-1}} \right)$ tends to zero as n tends to infinity.

The longer formulation is still useful for clarity, however, and we might need it if different terms are specified differently. For instance, the sequence $1, 0, 1, 0, 1, 0, \ldots$ could be specified like this:

Let (x_n) be the sequence defined by $x_n = \begin{cases} 1 \text{ if } n \text{ is odd} \\ 0 \text{ if } n \text{ is even} \end{cases}$.

This is just one sequence, so don't be tempted to think of it as 'two sequences' because of the way it is written. The formula gives a single value for each of x_1, x_2, x_3 and so on as usual.

Here are two more sequences, represented using both formulas and lists. Which formula goes with which list?

$1, 1, 2, 2, 3, 3, 4, 4, \ldots$ $\qquad\qquad$ $1, 3, 2, 4, 3, 5, 4, 6, \ldots$

$b_n = \begin{cases} (n+1)/2 \text{ if } n \text{ is odd} \\ n/2 \qquad \text{ if } n \text{ is even} \end{cases}$ $\qquad$ $c_n = \begin{cases} (n+1)/2 \text{ if } n \text{ is odd} \\ (n+4)/2 \text{ if } n \text{ is even} \end{cases}$

Formulas can be useful because they give abbreviated expressions for whole sequences. But I'd advise against getting obsessed with them. Translating between representations is an important skill, but students sometimes spend a long time worrying about how to write a formula when a list would get their point across perfectly well.

Sequences can also be represented graphically. One standard graphical representation is a number line, and for some sequences, such as $1, \frac{1}{2}, \frac{1}{4}, \frac{1}{8}, \frac{1}{16}, \ldots$, this works quite well:

Notice, though, that this diagram does not explicitly represent the order of the terms, so to 'read' it we have to impose some extra knowledge about which label refers to the first term, which to the second term, and so on. This makes number lines pretty useless for a sequence like $1, 0, 1, 0, 1, 0, \ldots$, although we could add accompanying labels to explain what is going on:

An alternative is to use an extra dimension, graphing a_n against n:

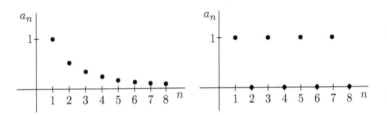

It is appropriate to use dots rather than curves for sequence graphs because each sequence is defined only for natural number values; there is no $a_{3/2}$, for instance. Notice also that this kind of graph uses an axis to explicitly represent the n values as well as the a_n values, so it gives a sense of the long-term behaviour of the sequence. That's handy because the long-term behaviour is often what we're interested in. What would graphs look like for the sequences (b_n) and (c_n) defined as earlier?

Graphs are also useful for thinking about a mathematical link to the concept of function: a sequence is technically *a function from the natural numbers to the reals*. Indeed, it might be defined as such at the beginning of an Analysis course. This probably sounds a bit unnatural compared with thinking of a sequence as an infinite list, but you should be able to see why it is reasonable by looking at the graph and by considering that every element of $\mathbb{N} = \{1, 2, 3, 4, 5, \ldots\}$ has a corresponding sequence term: the number 1 maps to a_1, the number 2 maps to a_2, etc. The link with functions would probably be clearer if instead of a_1 we wrote $a(1)$ or $f(1)$, but the subscript notation is standard for sequences. The link is worth noting, though—it is always valuable to see relationships across mathematical domains because theories developed for one concept might also apply to another.

5.3 Sequence properties: monotonicity

The various representations listed above can be useful for thinking about sequence properties. For instance, a sequence might be *increasing*, or *decreasing*, or *bounded* or *convergent*. What do you think these words mean? How would you explain their meanings to someone else? How would you formulate corresponding mathematical definitions using appropriate notation? Look away from the book and try this now.

If you gave that a serious go, it should be obvious that although your intuitive understanding might feel strong, it can be challenging to capture it in a coherent sentence. Awareness of this should put you in the right frame of mind for serious study of the definitions formulated by mathematicians.

Here are the definitions for *increasing* and for *decreasing*.

Definition: A sequence (a_n) is *increasing* if and only if $\forall n \in \mathbb{N}$,
$$a_{n+1} \geq a_n.$$

Definition: A sequence (a_n) is *decreasing* if and only if $\forall n \in \mathbb{N}$,
$$a_{n+1} \leq a_n.$$

These sound straightforward, but it is surprisingly difficult to think about how they combine. To see what I mean, consider these sequences. Would you say that each one is increasing, decreasing, both, or neither?

$1, 0, 1, 0, 1, 0, 1, 0, \ldots$

$1, 4, 9, 16, 25, 36, 49, \ldots$

$1, \frac{1}{2}, \frac{1}{3}, \frac{1}{4}, \frac{1}{5}, \frac{1}{6}, \frac{1}{7}, \frac{1}{8}, \ldots$

$1, -1, 2, -2, 3, -3, \ldots$

$3, 3, 3, 3, 3, 3, 3, 3, \ldots$

$1, 3, 2, 4, 3, 5, 4, 6, \ldots$

$6, 6, 7, 7, 8, 8, 9, 9, \ldots$

$0, 1, 0, 2, 0, 3, 0, 4, \ldots$

$10\frac{1}{2}, 10\frac{3}{4}, 10\frac{7}{8}, 10\frac{15}{16}, \ldots$

$-2, -4, -6, -8, -10, \ldots$

Almost everyone gets some of these wrong. So have another look, checking carefully against the definitions.

Here are the answers.

$1, 0, 1, 0, 1, 0, 1, 0, \ldots$	*neither*
$1, 4, 9, 16, 25, 36, 49, \ldots$	*increasing*
$1, \frac{1}{2}, \frac{1}{3}, \frac{1}{4}, \frac{1}{5}, \frac{1}{6}, \frac{1}{7}, \frac{1}{8}, \ldots$	*decreasing*
$1, -1, 2, -2, 3, -3, \ldots$	*neither*
$3, 3, 3, 3, 3, 3, 3, 3, \ldots$	*both*
$1, 3, 2, 4, 3, 5, 4, 6, \ldots$	*neither*
$6, 6, 7, 7, 8, 8, 9, 9, \ldots$	*increasing*
$0, 1, 0, 2, 0, 3, 0, 4, \ldots$	*neither*
$10\frac{1}{2}, 10\frac{3}{4}, 10\frac{7}{8}, 10\frac{15}{16}, \ldots$	*increasing*
$-2, -4, -6, -8, -10, \ldots$	*decreasing*

Were you right? Even when told to be careful, many Analysis students get at least one wrong. Most, for instance, want to classify $1, 0, 1, 0, 1, 0, 1, 0, \ldots$ as both increasing and decreasing, and almost everyone wants to classify $3, 3, 3, 3, 3, 3, 3, 3, \ldots$ as neither increasing nor decreasing. This is not surprising because these are perfectly natural interpretations. But they are based on everyday intuition, not on the mathematical definitions.

To understand the first case it helps to think about local versus global properties. When people say that the sequence $1, 0, 1, 0, 1, 0, 1, 0, \ldots$ is both increasing and decreasing, they are usually thinking about local properties. They see the sequence as starting at 1, then decreasing, then increasing, then decreasing, then increasing, and so on. But they should be thinking about a global property, because the definition of *increasing* is a universal statement: it says that for *every* $n \in \mathbb{N}$, $a_{n+1} \geq a_n$. That certainly isn't true for this sequence. Indeed it fails rather badly. There are infinitely many values of n for which a_{n+1} is not greater than or equal to a_n. For instance, $a_2 < a_1$, and $a_4 < a_3$, and so on. So this sequence does not satisfy the definition of increasing. Similarly, it does not satisfy the definition of decreasing. So, mathematically speaking, it is neither increasing nor decreasing.

To understand the second case it is necessary to be careful about the inequality. To satisfy the definition of *increasing*, each term must be greater than or equal to its predecessor. If every term is equal to its predecessor, that's enough. This might seem weird, but the definition is reasonable because it is simple and because it means that sequences like $6, 6, 7, 7, 8, 8, 9, 9, \dots$ get classified as increasing. It also works well within the theory of Analysis, because it lends itself to simple theorem statements—many theorems that apply to increasing sequences in general apply to constant sequences in particular. That said, mathematicians also use these definitions:

Definition: A sequence (a_n) is *strictly increasing* if and only if $\forall n \in \mathbb{N}$, $a_{n+1} > a_n$.

Definition: A sequence (a_n) is *strictly decreasing* if and only if $\forall n \in \mathbb{N}$, $a_{n+1} < a_n$.

Think about how these apply to the listed sequences too.

The final thing to know about the properties *increasing* and *decreasing* is that they are also associated with this definition:

Definition: A sequence (a_n) is *monotonic*[3] if and only if it is increasing or decreasing.

Students sometimes get confused about this because of the word 'or'. In everyday English, 'or' has two distinct meanings[4] and we are adept at using context and emphasis to work out which is intended. One meaning is *inclusive*, and is used when we mean one thing or the other or both, as in:

Students wishing to study Applied Statistics in year 3 should ensure that they take Statistical Methods or Introduction to Mathematical Statistics in year 2.

---

[3] People sometimes use the word *monotone* instead of *monotonic*.

[4] This is not the same in all languages—some have different words for inclusive and exclusive or.

The other meaning is *exclusive*, and is used when we mean one thing or the other but not both, as in:

Your lunch voucher entitles you to an ice cream or a slice of cake.

To avoid ambiguity in mathematics, we choose one meaning and stick to it, and the interpretation we use is the inclusive one. So this definition means that a sequence is monotonic if it is increasing or decreasing or both and, from the list, these sequences are classified as monotonic:

$1, 4, 9, 16, 25, 36, 49, \ldots$

$1, \frac{1}{2}, \frac{1}{3}, \frac{1}{4}, \frac{1}{5}, \frac{1}{6}, \frac{1}{7}, \frac{1}{8}, \ldots$

$3, 3, 3, 3, 3, 3, 3, 3, \ldots$

$6, 6, 7, 7, 8, 8, 9, 9, \ldots$

$10\frac{1}{2}, 10\frac{3}{4}, 10\frac{7}{8}, 10\frac{15}{16}, \ldots$

$-2, -4, -6, -8, -10, \ldots$

5.4 Sequence properties: boundedness and convergence

The definition of *bounded above* for a sequence is analogous to the definition of *bounded above* for a set, which was discussed in Section 2.6:

Definition: The set X is *bounded above* if and only if $\exists M \in \mathbb{R}$ such that $\forall x \in X, x \leq M$.

Definition: The sequence (a_n) is *bounded above* if and only if $\exists M \in \mathbb{R}$ such that $\forall n \in \mathbb{N}, a_n \leq M$.

The only difference is that '$\forall n \in \mathbb{N}$' is used in the sequence case because we always *index* the terms of a sequence using the natural numbers. I like graphical representations for thinking about boundedness. Look at these and make sure you can see how they relate to the definition:

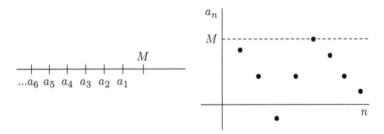

What do you think is the definition of *bounded below* and how would you represent the idea graphically?

Mathematicians also use another related definition, which I will simply present with accompanying diagrams.

Definition: The sequence (a_n) is *bounded* if and only if $\exists M > 0$ such that $\forall n \in \mathbb{N}, |a_n| \leq M$.

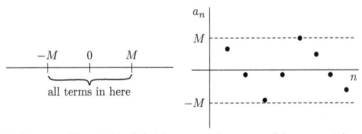

Is there anything in this definition to say that some of the terms either have to be or cannot be equal to M or $-M$? And why does it make sense to specify that $M > 0$?

Because graphical representations are useful for thinking about sequence properties, they can also be useful for understanding related theorems. I'd like to start considering theorems now, and I want to use the property *convergence*. But I will not introduce the definition of convergence yet—it is logically complex so I will devote separate sections to it later. For the time being, here is an informal description of its meaning, together with a diagram.

Informal description: A sequence (a_n) converges to a limit a if and only if, by going far enough along the sequence, we can make a_n as close as we like to a.

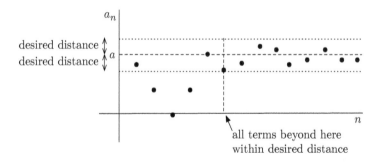

all terms beyond here
within desired distance

Note that if the desired distance were smaller, we might have to go further along the sequence. This description probably corresponds pretty well to your intuitive idea of convergence, but it might differ in a couple of respects. First, the everyday use of the word 'converges' tends to make people think only about monotonic sequences, and to believe that the terms must get closer and closer to the limit a in a rather simple way, like this:

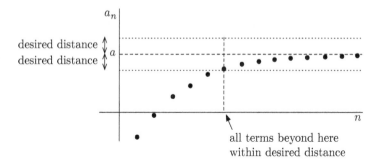

all terms beyond here
within desired distance

The informal description certainly applies to such sequences. But it also applies to sequences like that in the first diagram, which is more generic[5] in the sense that some terms are above the limit and some are below, and that sometimes the terms go away from the limit for a bit before getting closer again. In these respects, the informal description is faithful to the

[5] See the discussions about diagrams in Sections 2.5 and 2.9.

mathematical version of the concept, so you should adjust your thinking in that direction now.

Even with this fairly small range of properties, we can consider an array of possible theorems, some of which appear below. These would be referred to as *universal statements* because they each make a claim about every object that satisfies some properties. Which do you think are true and which do you think are false?

- Every bounded sequence is convergent.
- Every convergent sequence is bounded.
- Every monotonic sequence is convergent.
- Every convergent sequence is monotonic.
- Every monotonic sequence is bounded.
- Every bounded sequence is monotonic.
- Every bounded monotonic sequence is convergent.

To prove that a universal statement is false—to *refute* the statement—a mathematician would simply provide a *counterexample*. A counterexample to the first statement, for instance, would be a bounded sequence that is not convergent. One counterexample will do: if there exists a single example that does not satisfy a universal statement, that is enough to show that the statement is false. For those statements that you think are false, can you come up with specific counterexamples?

To prove that a universal statement is true, we have to prove that the conclusion really does hold for every object that satisfies the premises. Obviously this might involve more work than finding a single counterexample. Proofs for statements like these build fairly directly on the relevant definitions so we will consider some of them later in the chapter. For now, for those that you think are true, can you give a convincing intuitive argument that you are right?

We can formulate more theorems if we consider *subsequences*. A subsequence is exactly what it sounds like it should be: from an original sequence, we simply select some terms and leave out others. So, for example, the sequence

$$(a_n) = 1, \tfrac{1}{2}, \tfrac{1}{3}, \tfrac{1}{4}, \tfrac{1}{5}, \ldots$$

has subsequences including

$$(a_{2^n}) = \tfrac{1}{2}, \tfrac{1}{4}, \tfrac{1}{8}, \tfrac{1}{16}, \tfrac{1}{32}, \ldots \quad \text{and} \quad (a_{3n-1}) = \tfrac{1}{2}, \tfrac{1}{5}, \tfrac{1}{8}, \tfrac{1}{11}, \tfrac{1}{14}, \ldots.$$

Make sure you can see how the notation is used by substituting in $n = 1$, $n = 2$ and so on. When constructing a subsequence, messing around with the order of the terms is not allowed—otherwise the link to the original sequence would be lost. And stopping is not allowed—a subsequence has to be a sequence in its own right, so it must be infinite. Also, the subsequences listed above happen to follow algebraically expressible patterns regarding which terms are selected and which are not, but that is not necessary—we could create a subsequence by tossing a coin to decide whether or not each term gets included.

Here are some more universal statements that might be theorems. Which do you think are true and which do you think are false?

- Every convergent sequence has a monotonic subsequence.
- Every sequence has a monotonic subsequence.
- Every bounded sequence has a convergent subsequence.

If you found that easy to answer, then you haven't thought hard enough. If you're not convinced of this, you might like to know that when I asked a class of 200 Analysis students whether they thought the middle one was true, about half said yes and half said no. And that wasn't because they were giving ill-thought-out answers—they had worked with the definitions and they'd been given time to discuss what they thought. So, whatever you think, large numbers of smart and well-informed people would disagree.

With that in mind, have another go, remembering that a sequence does not have to follow a predictable pattern (though you might like to start by considering sequences from the list in Section 5.3). If you think that a statement is false, can you provide a specific counterexample? Or can you at least describe what a counterexample would be like? If you think that a statement is true, how would you convince someone else? If that person believed that there must be a counterexample, how would you convince them that they were wrong? Thinking about these things in detail will help you to appreciate the arguments you'll see later in this chapter and in an Analysis course.

5.5 Convergence: intuition first

This section and the next both address the definition of convergence. This one formalizes the informal description from Section 5.4. The next starts with the definition and explains how to understand it. One or other of these approaches might suit you better, so you might want to read these sections in reverse order. In particular, if you've already started an Analysis course and you're not understanding the longer definitions very well, you might find Section 5.6 useful because it incorporates general advice on how to work with such statements.

If you haven't yet started an Analysis course, you might be wondering why I'm making such a big deal of this. There are two reasons. One is that this definition is absolutely central in any Analysis course. The other is that it is the most logically complex definition to appear in work on sequences, so understanding it involves a bit of work.

Here, again, is the informal description:

Informal description: A sequence (a_n) converges to a limit a if and only if, by going far enough along the sequence, we can make a_n as close as we like to a.

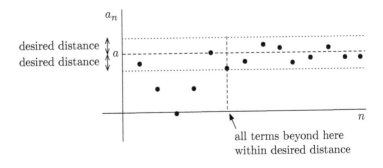

To convert this into a formal definition we need to get an algebraic handle on the idea of 'close'. Suppose we want to consider terms within distance ε of the limit a ('ε' is the Greek letter *epsilon*). In other words, we want terms between $a - \varepsilon$ and $a + \varepsilon$, so we want $a - \varepsilon < a_n < a + \varepsilon$.

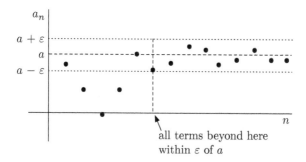

all terms beyond here
within ε of a

The inequalities $a - \varepsilon < a_n < a + \varepsilon$ can be written in the more abbreviated form $|a_n - a| < \varepsilon$, because

$$|a_n - a| < \varepsilon \Leftrightarrow -\varepsilon < a_n - a < \varepsilon$$
$$\Leftrightarrow a - \varepsilon < a_n < a + \varepsilon.$$

In this context I always read $|a_n - a| < \varepsilon$ as 'the distance between a_n and a is less than ε'. To keep track of the meaning, I advise you to do the same (and do read 'ε' as 'epsilon' and not as 'e').

Now, the whole description says '...by going far enough along the sequence, we can make a_n as close as we like to a.' Mathematicians capture this notion of 'far enough along' by saying

$$\exists N \in \mathbb{N} \text{ such that } \forall n > N, |a_n - a| < \varepsilon.$$

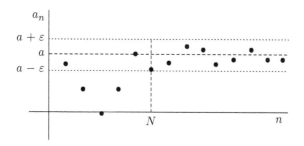

Don't forget to read the symbolic sentence out loud and think about how each part relates to the diagram. I think about it like this:

$\exists N \in \mathbb{N}$	such that	$\forall n > N,$	$|a_n - a| < \varepsilon.$
there is a point	such	beyond that	all the terms are
in the sequence	that	point	within epsilon of a.

This is only for one value of ε, however. Imagining a small ε captures the notion of making a_n close to a. But it doesn't capture the idea of making it *as close as we like*. We want to be able to make the terms within $\varepsilon = \frac{1}{2}$ of a, and within $\varepsilon = \frac{1}{4}$ of a, and so on, perhaps by going further along the sequence for smaller values of ε:

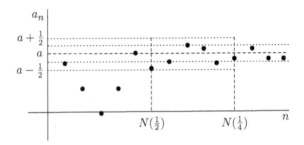

So in fact we want to say that for *any* $\varepsilon > 0$, it's possible to go far enough along the sequence to ensure that all the later terms are within distance ε of a. This leads to the whole definition:

Definition: (a_n) *converges to a if and only if*

$$\forall \varepsilon > 0 \; \exists N \in \mathbb{N} \text{ such that } \forall n > N, \; |a_n - a| < \varepsilon.$$

If you find it helpful to continue the informal wordy thinking, you could think of it this way:

Definition: (a_n) *converges to a if and only if*

$\forall \varepsilon > 0$	$\exists N \in \mathbb{N}$	such that	$\forall n > N,$	$|a_n - a| < \varepsilon.$
however small	there is a point	such	beyond that	all the terms are
epsilon is	in the sequence	that	point	within epsilon of a.

Don't be tempted to write down only a woolly, informal version, though. Mathematicians might think intuitively but they produce final written work using proper definitions.

5.6 Convergence: definition first

This section starts with the definition of convergence to a limit a. It explains how to understand this definition, in the process discussing ways to tackle similar statements. My aim is that you learn to look at a definition like this and work out what it means by linking it to other representations, so we'll end up where we ended up in the previous section—we'll just get there via a different route. Throughout, you should aim to understand why this is a reasonable definition of convergence.

Here is the definition:

Definition: (a_n) *converges to* a if and only if

$$\forall \varepsilon > 0 \, \exists N \in \mathbb{N} \text{ such that } \forall n > N, \, |a_n - a| < \varepsilon.$$

The first thing to try is reading this aloud ('ε' is the Greek letter *epsilon*, and if you've read Part 1 you'll probably remember the rest—if not, look at the symbol list in the Symbols section at the start of the book, on page xiii). Reading aloud is unlikely to give you an immediate sense of understanding, though, because everyday sentences are never this complicated, and neither are sentences in earlier mathematics. Specifically, this definition has three nested quantifiers—three of the quantifiers '$\forall$' and '$\exists$' piled up in one sentence. To understand a quantified sentence with a nested structure it is often easier to start at the end rather than the beginning, so we'll do that.

The last bit of the definition says '$|a_n - a| < \varepsilon$'. To see what this means it helps to do some algebra. Remember that, for instance,

$$|x| < 2 \Leftrightarrow -2 < x < 2.$$

By analogy, here we have

$$|a_n - a| < \varepsilon \Leftrightarrow -\varepsilon < a_n - a < \varepsilon$$
$$\Leftrightarrow a - \varepsilon < a_n < a + \varepsilon.$$

So $|a_n - a| < \varepsilon$ means that a_n is between $a - \varepsilon$ and $a + \varepsilon$. Or, if you prefer, that the distance between the term a_n and the limit a is less than ε. Because this involves comparing a_n and a, we can represent appropriate values on the vertical axis of a graph of a_n against n:

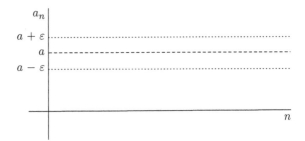

Working backwards through the definition, we next get '$\forall n > N$, $|a_n - a| < \varepsilon$'. In other words, for values of n bigger than N, the distance inequality holds. Note that we don't know anything about the terms beyond a_N except that they are within ε of a. And we don't know anything at all about the terms before a_N, so I won't put any there for now.

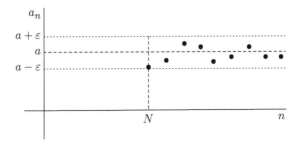

Taking another step back gives

$$\exists N \in \mathbb{N} \text{ such that } \forall n > N, \ |a_n - a| < \varepsilon.$$

In a sense, that's already represented on the diagram because N had to exist for me to draw it. But we might now want to think about the terms before a_N. Bothering to say that there exists an N beyond which something holds would make a mathematician think that before N, it might not hold—in this case, that some earlier terms might be further away from a. A generic diagram might therefore look something like this:

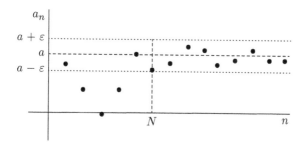

What about the remaining bit?

$$\forall \varepsilon > 0 \; \exists N \in \mathbb{N} \text{ such that } \forall n > N, \; |a_n - a| < \varepsilon.$$

This says that *for all epsilon greater than* 0, the stuff we've already looked at is true. Now, it makes sense to specify that $\varepsilon > 0$ because ε is a distance. And certainly the ε on the diagram above is greater than 0. But the diagram at the moment shows only one value of epsilon with its corresponding N. To understand what this means *for all epsilon greater than* 0, we could imagine allowing epsilon to vary, and consider that for smaller values of ε, we might need bigger values of N:

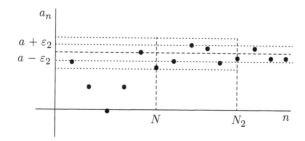

Overall, I like to capture the relationship between the informal and formal ideas by thinking of the definition, the diagram, and this informal interpretation:

Definition: (a_n) *converges to* a if and only if

$\forall \varepsilon > 0$	$\exists N \in \mathbb{N}$	such that	$\forall n > N,$	$|a_n - a| < \varepsilon.$
however small	there is a point	such	beyond that	all the terms are
epsilon is	in the sequence	that	point	within epsilon of a.

I hope you're now convinced that the definition sensibly captures the notion of convergence to a limit a. I also hope you found this explanation helpful. If you did, I'm pleased. However, there is a problem with helpful explanations: if you are reading or listening to one, it is easy to nod along thinking 'Yeah, yeah, yeah … yeah, I get that, okay. …' But that experience can be ephemeral—feeling that you have a good understanding at the time is not the same as being able to recall and apply that understanding later. If you want a more stringent test of whether or not you understood this section, write the definition on a blank piece of paper, put the book aside, and reconstruct the explanation for yourself.

5.7 Things to remember about convergence

The following sections demonstrate ways in which the definition of convergence is used to establish results in Analysis. Before that, though, I want to point out a few relationships between the definition and common intuitive ideas.

First, the explanations in the previous two sections do not 'prove' the definition. We don't prove definitions because they are just conventions: precise statements of meaning that everyone agrees to adopt. I've explained why the definition is reasonable by relating it to diagrams and to informal expressions, but that's not the same as proving it (see Sections 2.3 and 3.2 for information on definitions and on how they fit into mathematical theories).

Second, the definition doesn't say what happens 'at infinity'. This is good because, tempting though it is, talking about what happens 'at infinity' doesn't really make sense—there is no 'a_∞' or 'last term' of a sequence, because ∞ (infinity) is not a natural number.

Third, the definition requires no sense of motion or time. Many people think of convergence in terms of travelling along the sequence and watching the terms get closer to the limit a. But when mathematicians work with sequences, they do not imagine writing down new terms in a process that takes time. Instead, they treat the whole sequence as though it's 'already there'. Also, the image of moving along and watching as the terms get closer to a is a bit simplistic, because the definition does not specify that each term has to be closer to the limit than its predecessor. As noted in Section 5.4, that might be the case, but it might not.

Fourth, many people think of n as controlling a_n and therefore controlling the distance between a_n and a. That's fine, but when working with the definition we do not say 'for this N, this is the distance ε'. Instead we say 'for this distance ε, this is the appropriate N'. Look back to make sure you can see this, because it will be important for understanding how the definition is applied.

Fifth, several different notations and phrases are used when talking about convergence, and people tend to switch between them depending on what sounds more natural in a particular sentence. Some common ones are read aloud as shown here:

$(a_n) \to a$ '(a_n) converges to a' or '(a_n) tends to a'

$a_n \to a$ as $n \to \infty$ 'a_n tends to a as n tends to infinity'

$\lim\limits_{n \to \infty} a_n = a$ 'the limit as n tends to infinity of a_n is a'

In everyday English, these phrases might mean slightly different things. In mathematics, they all mean exactly the same thing: that (a_n) satisfies the definition of convergence.

Finally, I didn't pull the symbol 'ε' out of nowhere—everyone uses it in definitions associated with limits. The symbol 'ε' is like a small back-to-front '3', and is different from the set inclusion symbol '$\in$', which is more like a 'c' with an extra line. It is a good idea to make 'ε' and '$\in$' distinguishable in your handwriting because they often appear in the same sentence. You might also like to know that the resemblance between 'ε' and the more familiar symbol '3' means that every now and then students start the definition of convergence by writing '$\forall 3 > 0$'. This always makes me chuckle, but it's probably best to avoid it.

5.8 Proving that a sequence converges

After the definition of convergence is introduced, a lecturer will usually prove that certain sequences converge. In many cases, you'll be able to eyeball a sequence and identify its limit straightaway. So this is a situation in which we prove a result less in order to establish its truth and more in order to see how everything fits into a definition-based theory. Here we will consider the sequence (a_n) given by $a_n = 3 - \frac{4}{n}$ $\forall n \in \mathbb{N}$. Writing out the first few terms gives

$$\left(3 - \tfrac{4}{n}\right) = 3 - 4, \; 3 - 2, \; 3 - \tfrac{4}{3}, \; 3 - 1, \; 3 - \tfrac{4}{5}, \; 3 - \tfrac{4}{6}, \; 3 - \tfrac{4}{7}, \; 3 - \tfrac{4}{8}, \ldots.$$

What does this converge to? As n tends to infinity, $\frac{4}{n}$ gets really small, so a_n converges to 3.

To prove this we need to prove that the sequence satisfies the definition of convergence to 3; substituting in the appropriate values of a_n and a, this means we need to prove that

$$\forall \varepsilon > 0 \; \exists N \in \mathbb{N} \text{ such that } \forall n > N, \left|\left(3 - \tfrac{4}{n}\right) - 3\right| < \varepsilon.$$

People take different approaches to this, and either you or your lecturer might prefer to think about it purely logically and algebraically. As you know, however, I like diagrams, so I tend to start by sketching one. Here is a diagram showing the sequence (a_n) and an arbitrary-looking distance ε.

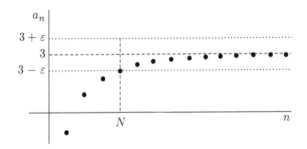

For a given value of ε, what value of N will do the job? If you can't answer immediately, ask yourself, what if ε were 1? What if ε were $\frac{1}{2}$, etc.? Clearly N is contingent upon ε: smaller values of ε require larger values of N. In general we want $\frac{4}{n} < \varepsilon$, which is equivalent to $\frac{4}{\varepsilon} < n$. So any natural number $N > \frac{4}{\varepsilon}$ will do. To give a specific one, mathematicians sometimes write $N = \left\lceil \frac{4}{\varepsilon} \right\rceil$, where $\lceil x \rceil$ is called the 'ceiling' of x and means the smallest integer larger than x. Having established this, we can use the definition as a guide to writing a proof. Here, again, is exactly what we need to prove:

$$\forall \varepsilon > 0 \, \exists N \in \mathbb{N} \text{ such that } \forall n > N, \left| \left(3 - \tfrac{4}{n}\right) - 3 \right| < \varepsilon.$$

We want to show that $\forall \varepsilon > 0$, something is true. So it makes sense to take an arbitrary $\varepsilon > 0$ and prove that the rest of the statement holds for this value, starting like this:

Claim: $\left(3 - \tfrac{4}{n}\right) \rightarrow 3$.
   Proof: Let $\varepsilon > 0$ be arbitrary.

For this ε, we want to show that there exists $N \in \mathbb{N}$ such that something is true. The easiest way to show that something exists is to specify what it should be, which we can do based on the reasoning above:

Claim: $\left(3 - \tfrac{4}{n}\right) \rightarrow 3$.
   Proof: Let $\varepsilon > 0$ be arbitrary.
      Set $N = \left\lceil \tfrac{4}{\varepsilon} \right\rceil$.

After that, we need to show that $\forall n > N, \left| \left(3 - \tfrac{4}{n}\right) - 3 \right| < \varepsilon$. This is straightforward, but make sure you can see why every equality and inequality is valid.

Claim: $\left(3 - \tfrac{4}{n}\right) \rightarrow 3$.
   Proof: Let $\varepsilon > 0$ be arbitrary.
      Set $N = \left\lceil \tfrac{4}{\varepsilon} \right\rceil$.
      Then $n > N \Rightarrow \left| \left(3 - \tfrac{4}{n}\right) - 3 \right| = \left| \tfrac{4}{n} \right| = \tfrac{4}{n} < \varepsilon$.

The proof is technically complete at this point because it establishes that the definition is satisfied. However, it is polite to write a conclusion. It would be fine simply to write 'Hence $\left(3 - \tfrac{4}{n}\right) \rightarrow 3$,' but you might also like to add an extra line to summarize the argument, especially if you think that will remind you how it all fits together:

Claim: $\left(3 - \frac{4}{n}\right) \to 3$.

Proof: Let $\varepsilon > 0$ be arbitrary.

Set $N = \left\lceil \frac{4}{\varepsilon} \right\rceil$.

Then $n > N \Rightarrow |a_n - 3| = \left|\left(3 - \frac{4}{n}\right) - 3\right| = \left|\frac{4}{n}\right| = \frac{4}{n} < \varepsilon$.

Hence we have shown that

$$\forall \varepsilon > 0 \; \exists N \in \mathbb{N} \text{ s.t. } \forall n > N, \left|\left(3 - \frac{4}{n}\right) - 3\right| < \varepsilon.$$

So $\left(3 - \frac{4}{n}\right) \to 3$ as required.

When studying a proof like this it is a good idea to think beyond the specifics, asking how it could be varied without damage to the main argument, or how it could be modified to deal with different cases. For instance, we used $N = \left\lceil \frac{4}{\varepsilon} \right\rceil$, but did we need to? Would $N = \left\lceil \frac{4}{\varepsilon} \right\rceil + 100$ do instead? Can you explain your answer by relating it both to the algebra and to a diagram? How would you modify this proof to deal with (a_n) given by $a_n = 3 + \frac{5}{n}$? How would you modify it to deal with (a_n) given by $a_n = c + \frac{d}{n}$, where c and d are constants? Would your modification work for both positive and negative values of c and d? For c and d that are equal to zero? For c and d that are not integers?

It's a good idea to do this because an Analysis course will include worked examples, but probably not many. A lecturer might, for instance, prove that $\left(\frac{1}{n}\right)$ converges to zero, then move straight on to another theoretical discussion. That might seem odd if you're accustomed to mathematics in which you're taught a procedure then asked to apply it lots of times (you might want to read or re-read Part 1 if so). But you should feel less need to see lots of examples or to do lots of practice if you use questions like these to think about generalizations.

5.9 Convergence and other properties

The previous section showed that a specific sequence satisfies the definition of convergence. This section will focus on proving things about relationships between convergence and other properties. Remember these possible theorems from Section 5.4?

- Every bounded sequence is convergent.
- Every convergent sequence is bounded.

What do you think? Is every bounded sequence convergent? No. Most people are okay with this, because they can easily think of some sequences that are bounded but not convergent. Here is one, for instance:

$$(x_n) \text{ given by } x_n = \begin{cases} 1 \text{ if } n \text{ is odd} \\ 0 \text{ if } n \text{ is even} \end{cases}.$$

Can you give another one? Can you give fifteen more? I'm joking, of course—but you should convince yourself that you easily could think of fifteen more, with at least some interesting variation.

What about the other statement? Is every convergent sequence bounded? Yes. But fewer people are okay with this, and here's the reason. Remember I said that at some point you'd be tempted to let a sequence be infinite 'in both directions'? This is where that temptation arises. People are accustomed to thinking about functions from the reals to the reals rather than about sequences—functions from the *naturals* to the reals. So they tend to either mentally or physically replace the dots appropriate to a sequence graph with a curve, and conjure up an image like this:

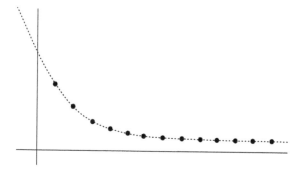

This makes them think that a convergent sequence could be unbounded 'to the left'. But a sequence is an infinite list of terms $a_1, a_2, a_3, \ldots$. The number a_1 is the first term—there is no 'to the left' of a_1. This is why I encourage students to sketch sequence graphs with dots only—it helps to circumvent this temptation.

So it is true that *every* convergent sequence is bounded, and this can be stated as a theorem. Here it is, with a proof and an accompanying diagram. Before you read this, you might want to review the self-explanation training in Section 3.5. As you read it, think in detail about the links between the proof and the labels on the diagram.

Theorem: Every convergent sequence is bounded.

    Proof: Suppose that $(a_n) \to a$.

        Then, by definition, $\exists N \in \mathbb{N}$ s.t. $\forall n > N$, $|a_n - a| < 1$,

                i.e. $\forall n > N$, $a - 1 < a_n < a + 1$.

        Note that N is finite.

        Let $M = \max\{|a_1|, |a_2|, \ldots, |a_N|, |a - 1|, |a + 1|\}$.

        Then $\forall n \in \mathbb{N}$, $|a_n| < M$.

        So (a_n) is bounded.

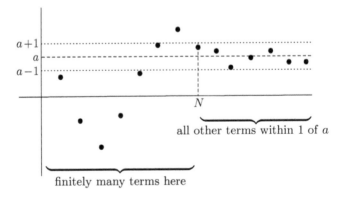

Where would M be on this diagram? Could you and would you draw or label the diagram differently?

One thing to notice here is that where the proof in the previous section concluded that the definition of convergence was satisfied, this proof starts by assuming that it is satisfied. That is appropriate to the structures of the claim and the theorem—make sure you can see how. Here, the assumption of convergence is used to deduce that

$\exists N \in \mathbb{N}$ such that $\forall n > N, |a_n - a| < 1$, which follows by simply replacing ε in the definition with the specific number 1 (the definition holds for all $\varepsilon > 0$, so it certainly holds for $\varepsilon = 1$).

A second thing to notice is that this true theorem has a false converse.[6] You might think about the information like this:

$$\text{convergent} \Rightarrow \text{bounded}$$
$$\text{bounded} \not\Rightarrow \text{convergent}$$

I wouldn't write this sort of thing anywhere official as it's very informal, but I find it quite useful in notes for myself.

Before we go on, here are the remaining statements from the original list. Do you now have better insight for some of the others, too?

- Every monotonic sequence is convergent.
- Every convergent sequence is monotonic.
- Every monotonic sequence is bounded.
- Every bounded sequence is monotonic.
- Every bounded monotonic sequence is convergent.

5.10 Combining convergent sequences

In Section 3.2 I noted that once a definition is introduced, early theorems often involve a claim that if two objects both satisfy that definition, then so does some combination of them. In this section we'll consider such a case.

Suppose that $(a_n) \to a$ and $(b_n) \to b$. What can we conclude about the sequence $(a_n + b_n)$? This is not a trick question. We can conclude that $(a_n + b_n) \to a + b$. This result (often called the *sum rule*) is obvious, so again the point of studying a proof is to understand how it can be proved within the formal theory. The proof invokes another theorem known as the triangle inequality (I will just state this here, but you should think about why it is reasonable and look out for a proof):

Theorem (the triangle inequality): $\forall x, y \in \mathbb{R}, |x + y| \leq |x| + |y|$.

---

[6] See Section 2.10 for a discussion of the technical meaning of *converse*.

The sum rule and a proof are given below. This proof is a classic, but it is a bit more complicated than the one in the previous section. Read it carefully, again applying the self-explanation training from Section 3.5. If you come to something you don't understand, try to articulate exactly what is puzzling you. After the proof, I will list some things that students commonly ask and provide answers.

Theorem (sum rule for convergent sequences):

Suppose that $(a_n) \to a$ and $(b_n) \to b$. Then $(a_n + b_n) \to a + b$.

Proof: Let $(a_n) \to a$ and $(b_n) \to b$.

Let $\varepsilon > 0$ be arbitrary.

Then $\exists N_1 \in \mathbb{N}$ s.t. $\forall n > N_1, |a_n - a| < \varepsilon/2$

and $\exists N_2 \in \mathbb{N}$ s.t. $\forall n > N_2, |b_n - b| < \varepsilon/2$.

Let $N = \max\{N_1, N_2\}$.

Then $\forall n > N$,

$$|(a_n + b_n) - (a + b)| = |a_n - a + b_n - b|$$
$$\leq |a_n - a| + |b_n - b|$$

by the triangle inequality

$$< \varepsilon/2 + \varepsilon/2$$
$$= \varepsilon.$$

So $\forall \varepsilon > 0 \; \exists N \in \mathbb{N}$ s.t. $\forall n > N, |(a_n + b_n) - (a + b)| < \varepsilon$.

So $(a_n + b_n) \to a + b$ as required.

Did you understand everything in that proof? Do you have questions in mind about anything you did not quite get? You don't want to rely on me unnecessarily, so, before you read on, imagine that you are explaining this theorem and proof to another person. Where, if anywhere, do you get stuck?

Here is a list of common student questions, together with answers.

- *Why start by setting $\varepsilon > 0$ to be arbitrary?*
  Because the conclusion, in the penultimate line, says that something is true for all $\varepsilon > 0$. Starting with an arbitrary $\varepsilon > 0$ means that the whole proof holds for any $\varepsilon > 0$.

- *Why say $|a_n - a| < \varepsilon/2$ instead of $|a_n - a| < \varepsilon$?*
  The answer to this lies in looking ahead and checking the algebra. Looking ahead, note that we want to end up with $|(a_n+b_n)-(a+b)|<\varepsilon$. Checking the algebra, note that this is done by adding together $|a_n-a|$ and $|b_n - b|$, so we want each of those to be less than $\varepsilon/2$.

- *But why is it okay to say that $\exists N_1 \in \mathbb{N}$ s.t. $\forall n > N_1$, $|a_n - a| < \varepsilon/2$?*
  Because if ε is an arbitrary number greater than 0, then $\varepsilon/2$ is just another number greater than 0. We are assuming that $(a_n) \to a$, so by definition there must exist $N \in \mathbb{N}$ such that $\forall n > N$, $|a_n - a|$ is less than this number $\varepsilon/2$. The proof uses N_1 as the name for such a number.

- *Why are the N-values called N_1 and N_2?*
  Because they might be different; the N-value beyond which $|a_n - a| < \varepsilon/2$ might not be the same as the N-value beyond which $|b_n - b| < \varepsilon/2$. Calling the numbers N_1 and N_2 is just a standard way of indicating that they don't have to be the same.

- *Why take the maximum of N_1 and N_2?*
  If we take $N = \max\{N_1, N_2\}$, then every $n > N$ will satisfy both $n > N_1$ and $n > N_2$. So for every $n > N$ we will have both $|a_n - a| < \varepsilon/2$ and $|b_n - b| < \varepsilon/2$, which is what we want.

This list of questions and answers encompasses a lot of reasoning that appears repeatedly in an Analysis course. You will see numerous proofs that involve judicious choice of an ε-value or using the triangle inequality to split up an expression. Because this book gives detail on only a few theorems, that might not be visible here. But look out for such tricks—students in my recent Analysis course told me that it got a lot easier once they started to see this repetition.

Look out in particular for similar ideas appearing in other theorems and proofs about combining sequences. For instance, the *product rule* states that if $(a_n) \to a$ and $(b_n) \to b$ then $(a_nb_n) \to ab$. Is this ringing any bells? The theorem that I used as an illustration in Chapter 1 is exactly this result. Its proof is more complicated again, and it involves two new tricks: adding and subtracting the same thing to make it easier to split up an expression, and adding 1 to the bottom of a fraction to ensure that we are not dividing by zero. With those things in mind, you should now be ready to turn back to Chapter 1 and have a go at reading it.

5.11 Sequences that tend to infinity

As well as considerable study of sequences that converge to finite limits, Analysis also involves study of sequences that tend to infinity. What do you think it means for a sequence to tend to infinity? Here is the definition, with some accompanying informal words and a diagram. Read everything properly and think about how you would explain this idea to someone else, perhaps using Sections 5.5 and 5.6 for inspiration.

Definition:[7] (a_n) *tends to infinity* if and only if

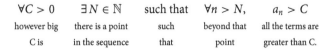

$\forall C > 0$	$\exists N \in \mathbb{N}$	such that	$\forall n > N,$	$a_n > C$
however big C is	there is a point in the sequence	such that	beyond that point	all the terms are greater than C.

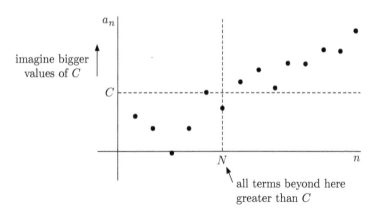

As with convergence, several different phrases are used in sentences about this concept:

$(a_n) \to \infty$	'(a_n) tends to infinity'
$a_n \to \infty$ as $n \to \infty$	'a_n tends to infinity as n tends to infinity'
$\lim_{n \to \infty} a_n = \infty$	'the limit as n tends to infinity of a_n is infinity'

[7] There is some variation in how people write this definition. It's perfectly fine, for instance, to start with '$\forall C \in \mathbb{R}$', but some people (including me) use '$\forall C > 0$' because it makes some of the proofs algebraically tidier. You might want to think about why it doesn't matter which we use.

In this case, the last one is less common. Some people think we shouldn't use it at all because ∞ is not a number so it makes no sense to say that something is equal to it. However, it is often notationally convenient to write about limits this way. Just be alert to the problem in case your lecturer is a stickler about this sort of thing.

It is worth pausing here to clarify the relationship between the idea of tending to infinity and a couple of different representations. Here, for instance, are two sequences that both tend to infinity:

$$(2^n) = 2, 4, 8, 16, 32, \ldots \qquad (3n - 1) = 2, 5, 8, 11, 14, \ldots$$

No one has trouble believing this—obviously both sequences get as big as you like. You might want to think about how to prove it, though. For an arbitrary value of C, what would N have to be to ensure that $\forall n > N$, $a_n > C$? How would you construct a proof?

Here is a sequence that doesn't tend to infinity:

$$\left(\tfrac{1}{n}\right) = 1, \tfrac{1}{2}, \tfrac{1}{3}, \tfrac{1}{4}, \tfrac{1}{5}, \ldots$$

No one has trouble believing this either, at least when looking at the sequence in list form—obviously it tends to zero. However, people do occasionally get confused when looking at graphs like the one below. They look at the terms and imagine extending the graph infinitely to the right, and consequently find themselves thinking that $\left(\tfrac{1}{n}\right)$ tends to infinity.

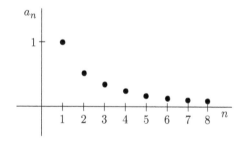

That must be nonsense, because if it weren't then all sequences would tend to infinity (they all extend infinitely to the right). To avoid falling into this trap, think about what is represented on which axis. We are talking about whether or not a_n tends to infinity, and a_n is represented on the vertical axis.

Combining sequences that tend to infinity opens up some interesting questions. For instance, these sequences both tend to infinity:

$$(n^2) = 1, 4, 9, 16, 25, \ldots \qquad (2^n) = 2, 4, 8, 16, 32, \ldots$$

What about this one?

$$\left(\frac{n^2}{2^n}\right) = \frac{1}{2}, \frac{4}{4}, \frac{9}{8}, \frac{16}{16}, \frac{25}{32}, \ldots$$

In this sequence, the numerator tends to infinity, but so does the denominator. Does one of them 'win', dominating the other and forcing the sequence to tend to zero or infinity? Or do they, perhaps, 'balance out' so that the sequence converges to 1, or maybe 2? Listing just the first few terms makes any of these outcomes seem plausible, but if you've read Section 2.9 and you are alert to links across mathematical ideas, you might immediately recognize what must happen. If not, look at that section and also try writing out a few more terms.

In some similar cases we can get better insight by thinking about the structures of the numerator and the denominator. For instance, consider the sequence

$$\left(\frac{6^n}{n!}\right) = \frac{6}{1}, \frac{36}{2}, \frac{216}{6}, \frac{1296}{24}, \frac{7776}{120}, \ldots$$

The numerator in this case involves some pretty big numbers, but in fact the sequence tends to zero. It's possible to see why by thinking about large values of n while keeping the structures visible:

$$\frac{6 \times 6 \times \ldots \times 6}{1 \times 2 \times \ldots \times n}.$$

If n is, say, 1000, then most of the multiplied numbers on the bottom are much bigger than their corresponding sixes on the top. That's a useful way to think about it intuitively; in an Analysis course you will likely learn about the *ratio test*, which provides a way to prove things like this formally. Here is a statement of the ratio test:

Theorem (ratio test):

Suppose that (a_n) is a sequence such that $(a_{n+1}/a_n) \to l$. Then:

1. If $-1 < l < 1$ then $(a_n) \to 0$.
2. If $l > 1$ and $a_n > 0$ $\forall n \in \mathbb{N}$ then $(a_n) \to \infty$.
3. If $l > 1$ and $a_n < 0$ $\forall n \in \mathbb{N}$ then $(a_n) \to -\infty$.
4. If $l < -1$ then the sequence neither converges nor tends to $\pm\infty$.
5. If $l = 1$ we get no information.

Why is this called the ratio test? Can you work out how to apply it to prove that both of the sequences

$$\left(\frac{n^2}{2^n} \right) \text{ and } \left(\frac{6^n}{n!} \right)$$

tend to zero (hint: what is the ratio a_{n+1}/a_n in each case)? Why does the ratio test apply so tidily to sequences like these? Can you come up with sequences that yield other limits of ratios and therefore lead to other conclusions? And why is each part of the theorem true? I won't prove the ratio test here (though see Section 6.6 for an analogous result about series). But a typical proof relies on several other theorems and thus constitutes a nice example of theory building—look out for it in your course.

As a final idea in this section, consider this theorem:

Theorem: If $(a_n) \to \infty$ then $\left(\dfrac{1}{a_n} \right) \to 0$.

This will also likely be proved in your course. Is its converse true?

This is another question that split my class of 200 down the middle: about half said yes and half said no, even after some discussion and an opportunity to rethink their answers. So think about whether you might have missed something, then read on.

Again the converse is not true. For *some* sequences that tend to zero, the reciprocal does tend to infinity. For example:

$$\left(\frac{1}{n}\right) \to 0 \text{ and } \left(\frac{1}{\frac{1}{n}}\right) = (n) \to \infty.$$

But this is not the case for all such sequences. For example:

$$\left(\frac{-1}{n}\right) \to 0 \text{ but } \left(\frac{1}{\frac{-1}{n}}\right) = (-n) \to -\infty.$$

There are worse cases too:

$$\left(\frac{(-1)^n}{n}\right) \to 0 \text{ but } \left(\frac{1}{\frac{(-1)^n}{n}}\right) = ((-1)^n n) = -1, 2, -3, 4, -5, 6, \ldots,$$

which does not tend to a limit of any kind.

If you got that wrong, don't worry about it. It's very common for people to think only about positive numbers, and to forget that different things might happen with negative ones (though you should learn from this and become more alert to similar problems). Indeed, if you have a reasonably creative Analysis lecturer then you will encounter numerous opportunities to be wrong—Analysis contains many theorems with plausible but false converses, and these make for great true/false questions to get people thinking in lectures and problems classes. In general, opportunities to be wrong are to be embraced. What I say to my classes is that I don't care who is right or wrong, I just care that everyone is thinking and is willing to change their minds if presented with a good reason to do so.

5.12 Looking ahead

The preceding sections are just a taster of the material you will study in an Analysis course that covers sequences. We haven't, for instance, looked at all the possible theorems in Section 5.4, and a typical course will involve proofs of all the true ones. It will also involve considerable work with 'standard' sequences like (x^n) and $(x^{\frac{1}{n}})$ and (n^α). Do these have limits as n tends to infinity? Does the answer differ for different values of x and α? And what happens for a sequence like $\left((3^n + 7^n)^{\frac{1}{n}}\right)$?

Then there are more general theorems that can be proved more or less directly from the definition of convergence. It is common, for instance, to prove that the limit of a sequence must be unique—that a sequence cannot tend to more than one limit. As with some other results in this chapter, this is completely obvious, so what you're supposed to learn is how to prove it within the formal theory. Similarly, you will probably see a proof of this theorem:

Theorem (sandwich rule):

Suppose that $(a_n) \to a$ and $(c_n) \to a$ and that $\forall n \in \mathbb{N}, a_n \leq b_n \leq c_n$. Then $(b_n) \to a$.

If you have understood the definition of convergence and you're willing to draw a few diagrams and/or do some algebra, you will probably be able to prove this now. And can you invent a *comparison test* along the same lines for sequences that tend to infinity? Such rules and tests, when combined with knowledge about a few standard sequences, can be used to establish the limiting behaviours of a great many more.

Many courses also involve study of *Cauchy*[8] *sequences*, the definition for which appears here:

Definition: (a_n) is a *Cauchy* sequence if and only if

$$\forall \varepsilon > 0 \, \exists N \in \mathbb{N} \text{ such that } \forall n, m > N, \, |a_n - a_m| < \varepsilon.$$

Where the definition of convergence is about terms getting close to a limit, the definition of a Cauchy sequence is about terms getting close to each other. Do you think that Cauchy sequences have to be convergent, and vice versa?

Finally, most courses on sequences go on to work on series. This is not a coincidence: sequence convergence is key to the theory of series, as I will explain in the next chapter.

---

[8] This is a French name, so the pronunciation is more like 'coe-shee' than 'cow-chee'.

CHAPTER 6

Series

This chapter starts with geometric series, examining conditions under which a formula can be used to calculate an infinite sum. It discusses notation, graphical representations and definitions for partial sums and series convergence, applying these to the harmonic series. It introduces tests for series convergence, describing some relationships between these, before demonstrating that infinite series can have some very peculiar behaviours. It concludes with a section on power series, followed by two sections on Taylor series and their relationships with functions.

6.1 What is a series?

A series is an infinite sum, like this:

$$1 + \tfrac{1}{3} + \tfrac{1}{9} + \tfrac{1}{27} + \tfrac{1}{81} + \dots.$$

As usual, the ellipsis at the end means 'and so on (forever)'. This series is a *geometric series* with common ratio $\tfrac{1}{3}$, and you might know that it therefore adds up to $\dfrac{1}{1 - \tfrac{1}{3}} = \tfrac{3}{2}$.

It's worth checking that this seems reasonable, even if you know the standard formula—people sometimes get good at working with formulas but forget to think about what they mean. In this case we could think about a number line:

Eyeballing this should convince you that the sum $\tfrac{3}{2}$ seems about right.

You might also know how the formula for the sum of a geometric series is derived, having seen an argument like the one below (in which I've rewritten $1 + \frac{1}{3} + \frac{1}{9} + \frac{1}{27} + \frac{1}{81} + \ldots$ as $1 + \frac{1}{3} + \frac{1}{3^2} + \frac{1}{3^3} + \frac{1}{3^4} + \ldots$ to make it easier to see what's going on).

Claim: $1 + \frac{1}{3} + \frac{1}{3^2} + \frac{1}{3^3} + \frac{1}{3^4} + \ldots = \frac{3}{2}.$

Proof: Let $$S = 1 + \frac{1}{3} + \frac{1}{3^2} + \frac{1}{3^3} + \frac{1}{3^4} + \ldots.$$

Then $$\frac{1}{3}S = \frac{1}{3} + \frac{1}{3^2} + \frac{1}{3^3} + \frac{1}{3^4} + \ldots.$$

So $$S - \frac{1}{3}S = 1,$$

i.e. $$\left(1 - \frac{1}{3}\right)S = 1.$$

So $$S = \frac{1}{1 - \frac{1}{3}} = \frac{3}{2}.$$

I think this is a lovely little proof. It finds the sum in an indirect way by giving it a name, S, then doing some multiplication that makes elegant use of the fact that the series has infinitely many terms, then making S the subject of a formula. It is also straightforward to generalize. If a geometric series has first term a and common ratio r, then an argument of the same form can be used to prove a more general theorem.

Theorem: $a + ar + ar^2 + ar^3 + ar^4 + \ldots = \dfrac{a}{1 - r}.$

Proof: Let $$S = a + ar + ar^2 + ar^3 + ar^4 + \ldots.$$

Then $$rS = ar + ar^2 + ar^3 + ar^4 + \ldots.$$

So $$S - rS = a,$$

i.e. $$(1 - r)S = a.$$

So $$S = \frac{a}{1 - r}.$$

As with much mathematics, the great thing is that this always works.

Or does it?

What if the first term is 1 and the common ratio is 3? Then the formula gives

$$1 + 3 + 9 + 27 + 81 + \ldots = \frac{1}{1 - 3} = \frac{1}{-2} = -\frac{1}{2}.$$

This, clearly, is nonsense. The infinite sum $1 + 3 + 9 + 27 + 81 + \ldots$ and the number $-\frac{1}{2}$ are very far from equal. The series does not even add up to a finite number, never mind a negative one.

And how about a common ratio of -1? The formula in this case gives

$$1 - 1 + 1 - 1 + 1 - \ldots = \frac{1}{1 - (-1)} = \frac{1}{2}.$$

But in fact this series does not add up to anything—the sum keeps alternating between 1 and 0. To say that this is somehow equal to $\frac{1}{2}$ is meaningless.

So the formula does not always work. Far from it.

A student new to advanced mathematics might not be alert to such problems because earlier mathematics often involves only problems or exercises for which a standard method applies. You probably 'knew' that the formula should only be applied if $|r| < 1$, because those are the only ratios for which you've been asked to use it. It doesn't take much thought to establish that it doesn't apply for all possible ratios. But this raises an interesting question and a point worth noting.

The interesting question is, what went wrong with the proof? It looked like it ought to work for every ratio, but it doesn't. There must be some hidden assumption in the reasoning, one that is not always valid. Did you get a sense of what that assumption might be? The problem occurs where we perform the subtraction $S - rS$. Notice that if $ar + ar^2 + ar^3 + ar^4 + \ldots = C$ then $S - rS = a + C - C$. This looks like it should be equal to a, and indeed it is if C happens to be a finite number. But if C is infinite, then we end up with '$a + \infty - \infty$', which is not a meaningful expression because '$\infty - \infty$' is not meaningful. Think, for instance, about subtracting the number of square numbers from the number of natural numbers (we would want the answer to be $+\infty$), then about subtracting the number of integers from the number of natural numbers (we would want the answer to be $-\infty$). Subtraction is *not well defined* in this case. Knowledge about finite objects does not necessarily generalize to infinite ones, and much of this chapter is about cases in which the mathematics of infinite series differs from that of finite sums.

The point worth noting is that advanced mathematics is less about performing calculations to find answers and more about pinning down the conditions under which a result or formula is valid. In the case of geometric series we might ask,

For what values of a and r is it true that $a + ar + ar^2 + ar^3 + \ldots = \dfrac{a}{1-r}$?

This type of question pervades the mathematics of series, and a typical Analysis course involves establishing tests that address the corresponding general question that can be asked about any series:

Does this add up to a finite number?

Section 6.4 will answer the question about geometric series, and later sections will introduce some of the available tests. First, however, we will set up some technical machinery to make the study of series easier.

6.2 Series notation

Series are not conceptually difficult, but they come wrapped in a lot of notation that makes them look complicated. This makes some students avoid questions about them on exams, when in fact those questions are often the easiest. You don't want to avoid easy exam questions, so it's worth getting to grips with the notation.

Many readers will already know that series can be represented using 'sigma notation', so called because it involves the upper-case Greek letter sigma, written 'Σ'. We write things like this:

$$\sum_{n=1}^{\infty} \frac{1}{3^{n-1}}$$ 'the sum from n equals one to infinity of one over three to the power n-minus-one'

To expand this notation, we take the values of n one at a time and add the corresponding terms together. So this is just another way of representing our original series.

$$\sum_{n=1}^{\infty} \frac{1}{3^{n-1}} = \frac{1}{3^{1-1}} + \frac{1}{3^{2-1}} + \frac{1}{3^{3-1}} + \ldots = 1 + \frac{1}{3} + \frac{1}{3^2} + \ldots$$

Sigma notation might seem cumbersome at first, but it has some big advantages. First, formulating a general expression for the terms of a series forces us to think about their structure. This might not buy us much for a simple geometric series, but it's handy for dealing with more complicated series like this:

$$\frac{1}{2} + 1 + \frac{9}{8} + 1 + \frac{25}{32} + \frac{18}{32} + \frac{49}{128} + \ldots = \sum_{n=1}^{\infty} \frac{n^2}{2^n}.$$

Second, we can vary the limits to express a series that 'starts' at a different place or to express a finite sum:

$$\sum_{n=5}^{\infty} \frac{n^2}{2^n} = \frac{5^2}{2^5} + \frac{6^2}{2^6} + \frac{7^2}{2^7} + \ldots;$$

$$\sum_{n=1}^{5} \frac{n^2}{2^n} = \frac{1^2}{2^1} + \frac{2^2}{2^2} + \frac{3^2}{2^3} + \frac{4^2}{2^4} + \frac{5^2}{2^5}.$$

A finite sum is not a series, of course, and there isn't much mileage in writing a finite sum in this way if it has only two or three terms. But it saves a bit of effort when expressing a sum with, say, ten related terms:

$$\frac{1}{1!} + \frac{1}{2!} + \frac{1}{3!} + \frac{1}{4!} + \frac{1}{5!} + \frac{1}{6!} + \frac{1}{7!} + \frac{1}{8!} + \frac{1}{9!} + \frac{1}{10!} = \sum_{n=1}^{10} \frac{1}{n!}.$$

There is also good reason to do it if we want to consider numerous related sums, like this:

Let $s_n = \sum_{i=1}^{n} \frac{1}{i}$. Then

$$s_1 = \sum_{i=1}^{1} \frac{1}{i} = \frac{1}{1}$$

$$s_2 = \sum_{i=1}^{2} \frac{1}{i} = \frac{1}{1} + \frac{1}{2}$$

$$s_3 = \sum_{i=1}^{3} \frac{1}{i} = \frac{1}{1} + \frac{1}{2} + \frac{1}{3}, \text{ etc., and in general}$$

$$s_n = \sum_{i=1}^{n} \frac{1}{i} = \frac{1}{1} + \frac{1}{2} + \frac{1}{3} + \ldots + \frac{1}{n}.$$

Notice that there are two variables in use here. One, i, is an *indexing* variable: in each term, we replace this with whatever number we're up to. The other, n, is a sort of stopping variable. Earlier in this section I used n as the indexing variable. That's fine as it is just a name—we can use whatever letter or symbol we like. But, in work with series, we sometimes care about both of these variables and it is useful to be able to distinguish them, so when I need both I will use i for indexing and n for stopping. I will also adopt the standard convention that, because we are mostly interested in infinite series, the notation $\sum a_n$ (with no specified limits) means the infinite series $a_1 + a_2 + a_3 + \ldots$.

In any case, sigma notation is very compact. For an expert, this is an advantage. But compact notation hides away much of the meaning, so students sometimes find themselves staring at an expression involving sigma notation as though it's just a meaningless jumble of symbols. I have a maxim for dealing with this situation:

If in doubt, write it out.

I realize this sounds a bit naff,[1] but the message is serious. If you are faced with a series (or a finite sum) written in sigma notation, writing out the first few terms will often give a better sense of what you're dealing with.

6.3 Partial sums and convergence

Recall that the general question asked about any series is:

Does this add up to a finite number?

We have seen an example that does, an example that doesn't because it has an infinite total, and an example that doesn't because it doesn't have a meaningful total at all:

$$\sum_{n=1}^{\infty} \frac{1}{3^{n-1}} \qquad \sum_{n=1}^{\infty} 3^n \qquad \sum_{n=1}^{\infty} (-1)^{n-1}$$

[1] This is a British-English word meaning, according to my computer's dictionary, 'lacking taste or style'. I lived in America for a bit but never did find an adequate American equivalent.

We have also seen examples for which the answer is less obvious. What about this series from the previous section, for instance?

$$\sum_{n=1}^{\infty} \frac{n^2}{2^n} = \frac{1^2}{2^1} + \frac{2^2}{2^2} + \frac{3^2}{2^3} + \frac{4^2}{2^4} + \frac{5^2}{2^5} + \dots$$

This has only positive terms, so its total must be either infinite or a finite positive number. Which do you think it is? Section 6.6 will provide the answer, but before that I want to make sure that you understand the potential complexity of the general question. To do so it is useful to introduce some notation, the notion of *partial sums*, and a graphical representation.

To talk about series in general I will use the notation

$$\sum_{n=1}^{\infty} a_n = a_1 + a_2 + a_3 + a_4 + \dots .$$

At this point it is worth explicitly contrasting series with sequences. In everyday English, people tend to use these words interchangeably. In mathematics, however, a sequence is an infinite *list*

$$(a_n) = a_1, a_2, a_3, a_4, a_5, a_6, \dots ,$$

whereas a series is a infinite *sum*

$$\sum a_n = a_1 + a_2 + a_3 + a_4 + a_5 + a_6 + \dots .$$

Obviously these are very different so it is important to use the language correctly. It is particularly important here because mathematicians relate each series to its sequence of *partial sums*.

Definition: The nth *partial sum* of the series $\sum_{i=1}^{\infty} a_i$ is $s_n = \sum_{i=1}^{n} a_i$.

This gives

$$s_1 = \sum_{i=1}^{1} a_i = a_1$$

$$s_2 = \sum_{i=1}^{2} a_i = a_1 + a_2$$

$$s_3 = \sum_{i=1}^{3} a_i = a_1 + a_2 + a_3, \text{ etc., and in general}$$

$$s_n = \sum_{i=1}^{n} a_i = a_1 + a_2 + a_3 + \ldots + a_n.$$

Why do you think these are called *partial sums*? This is not a trick question—I just want you to think about it so that you're not tempted to memorize the definition meaninglessly. And can you see why it is important to keep the language straight? The partial sums form a sequence (s_n), so each series has an associated sequence and it is important to know which notation is used for which.

The relationship becomes clearer, in my view, when looking at a graph showing s_n against n. The diagram below shows the graph for the given series, along with the first few partial sums. The graph uses dots rather than a curve because, as with sequences, the value s_n only exists when n is a natural number.

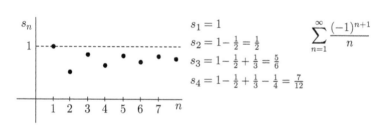

$$s_1 = 1$$
$$s_2 = 1 - \tfrac{1}{2} = \tfrac{1}{2}$$
$$\sum_{n=1}^{\infty} \frac{(-1)^{n+1}}{n}$$
$$s_3 = 1 - \tfrac{1}{2} + \tfrac{1}{3} = \tfrac{5}{6}$$
$$s_4 = 1 - \tfrac{1}{2} + \tfrac{1}{3} - \tfrac{1}{4} = \tfrac{7}{12}$$

Notice that a_n is not plotted explicitly on this graph, though we can 'see' it as the vertical difference between s_{n-1} and s_n. Notice also that the series has a finite sum if and only if the sequence (s_n) converges to a limit—look at the graph and think about why. This leads to the following definition.

Definition:

$$\sum_{i=1}^{\infty} a_i \text{ converges if and only if } (s_n) \text{ converges, where } s_n = \sum_{i=1}^{n} a_i.$$

The language gets a bit weird here, because we are really interested in whether or not the series adds up to a finite number. But, because the series is infinite, we approach this question via the sequence of partial sums. This leads us to describe the behaviour of the series in terms of convergence. You need to know the formal definition in terms of partial sums, obviously, but here is a summary of the conceptual information:

- We say that a series converges if it adds up to a finite number;
- We say that it diverges if it doesn't.

6.4 Geometric series again

Working with partial sums converts a question about series into a question about sequences. This allows us to give a precise answer to the earlier question about geometric series:

For what values of a and r is it true that $a + ar + ar^2 + ar^3 + \ldots = \dfrac{a}{1-r}$?

Using partial sums means that we can apply the familiar argument to the partial sum s_n, which is finite so that we don't run into problems with infinite or undefined sums. Then we can ask what happens to s_n as n tends to infinity, in effect turning the question about an infinite sum into a question about finite sums and a limit. Here is a whole argument, presented as a theorem and proof.

Theorem: $a + ar + ar^2 + ar^3 + ar^4 + \ldots = \dfrac{a}{1-r}$ if and only if $|r| < 1$.

Proof: Let $\qquad s_n = a + ar + ar^2 + \ldots + ar^{n-1}$.

Then $\qquad rs_n = \quad ar + ar^2 + \ldots + ar^{n-1} + ar^n$.

So $\quad s_n - rs_n = a - ar^n$,

i.e. $\quad (1-r)s_n = a - ar^n$.

So $\qquad s_n = \dfrac{a(1-r^n)}{1-r}$.

Now, (r^n) converges if and only if $|r| < 1$.

So (s_n) converges if and only if $|r| < 1$.

In such cases, $(r^n) \to 0$ so $(s_n) \to \dfrac{a}{1-r}$.

With this established, I think it is fun to look at visual representations for certain sums. For instance, imagine that the area of the whole square in the next diagram is 1. What is the area of the biggest black square? And the next biggest one? How does the picture illustrate the sum?

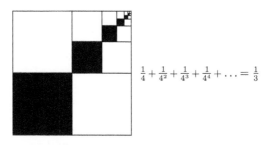

$$\frac{1}{4} + \frac{1}{4^2} + \frac{1}{4^3} + \frac{1}{4^4} + \ldots = \frac{1}{3}$$

Another image that you might have seen is the *Koch snowflake*, which is constructed iteratively by taking an equilateral triangle and adding on three triangles to construct a six-pointed star, then smaller triangles and so on. If the area of the original triangle is 1, what is the area of the star? And what is the area at the next iteration? What geometric series does this construction correspond to, and what is its sum (the area of the limiting shape)? And, for something a bit weirder, what is the perimeter of that shape?

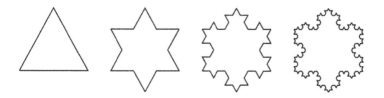

6.5 A surprising example

The proof for the geometric series formula used the fact that the sequence (r^n) tends to zero only if $|r| < 1$. For a general series $\sum a_n = a_1 + a_2 + a_3 + \ldots$ to converge, it should be pretty clear that

a_n will have to tend to zero as n tends to infinity. This is captured in the following theorem.

Theorem: If $\sum a_n$ converges then $(a_n) \to 0$.

This theorem is sometimes referred to as the *null sequence test*, because its contrapositive[2] acts as a test for non-convergence:

(Contrapositive) If $(a_n) \not\to 0$ then $\sum a_n$ does not converge.

What about the converse of the theorem?

(Converse) If $(a_n) \to 0$ then $\sum a_n$ converges.

Many people just assume that this is true, even if they are alert to the fact that a conditional statement and its converse are different (see Section 2.10). It is intuitively natural to think that if the terms tend to zero then the series must have a finite total. But in fact this is not true, as shown by the counterexample below. This really caught my attention when I first saw it—partly because the result surprised me and partly because I found the associated argument so elegant and convincing.

Consider the *harmonic series*

$$\sum_{i=1}^{\infty} \frac{1}{n} = 1 + \frac{1}{2} + \frac{1}{3} + \frac{1}{4} + \frac{1}{5} + \frac{1}{6} + \dots,$$

for which the first few partial sums and a graph are shown here:

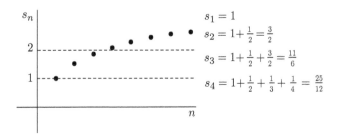

$s_1 = 1$

$s_2 = 1 + \frac{1}{2} = \frac{3}{2}$

$s_3 = 1 + \frac{1}{2} + \frac{3}{2} = \frac{11}{6}$

$s_4 = 1 + \frac{1}{2} + \frac{1}{3} + \frac{1}{4} = \frac{25}{12}$

---

[2] The *contrapositive* of the conditional statement 'if A then B' is 'if not B then not A.' If a conditional statement is true then its contrapositive is always true—see a transition-to-proof textbook or Section 4.6 of *How to Study for/as a Mathematics Degree/Major*.

Looking at this information for the first time, most people conclude that the series has a finite sum that is somewhere in the region of 3 to 5, or maybe 10 at the outside. But that is wrong. The sum is infinitely large, which we can demonstrate as follows.

The first term of the series, 1, is greater than $\frac{1}{2}$. The second is equal to $\frac{1}{2}$. The third term is not greater than or equal to $\frac{1}{2}$ on its own, but taking the next two terms gives something that is: $\frac{1}{3} + \frac{1}{4} > \frac{2}{4} = \frac{1}{2}$. Similarly, taking the next four terms, $\frac{1}{5} + \frac{1}{6} + \frac{1}{7} + \frac{1}{8} > \frac{4}{8} = \frac{1}{2}$. And we can add another half by taking the next eight terms, then the next sixteen, then the next thirty-two, and so on. Because we can keep adding more halves, the whole sum is infinite.

Sometimes people represent the argument like this:

$$1 + \tfrac{1}{2} + \underbrace{\tfrac{1}{3} + \tfrac{1}{4}}_{> \frac{1}{2}} + \underbrace{\tfrac{1}{5} + \tfrac{1}{6} + \tfrac{1}{7} + \tfrac{1}{8}}_{> \frac{1}{2}} + \underbrace{\tfrac{1}{9} + \tfrac{1}{10} + \tfrac{1}{11} + \tfrac{1}{12} + \tfrac{1}{13} + \tfrac{1}{14} + \tfrac{1}{15} + \tfrac{1}{16}}_{> \frac{1}{2}} + \cdots$$

To formalize it, a lecturer or textbook might use a claim like

$$\forall n \in \mathbb{N}, \ s_{2^n} > \frac{n+1}{2}.$$

Can you see why this is true? Again, the maxim 'if in doubt, write it out' applies. I would write out the inequality for a few values of n like this:

$$s_{2^1} = s_2 > \tfrac{1+1}{2} = 2 \times \tfrac{1}{2} \quad \text{because } s_2 = 1 + \tfrac{1}{2}$$

$$s_{2^2} = s_4 > \tfrac{2+1}{2} = 3 \times \tfrac{1}{2} \quad \text{because } s_4 = 1 + \tfrac{1}{2} + \underbrace{\tfrac{1}{3} + \tfrac{1}{4}}_{> \frac{1}{2}}$$

$$s_{2^3} = s_8 > \tfrac{3+1}{2} = 4 \times \tfrac{1}{2} \quad \text{because } s_8 = 1 + \tfrac{1}{2} + \underbrace{\tfrac{1}{3} + \tfrac{1}{4}}_{> \frac{1}{2}} + \underbrace{\tfrac{1}{5} + \tfrac{1}{6} + \tfrac{1}{7} + \tfrac{1}{8}}_{> \frac{1}{2}}$$

This convinces me that the claim is valid. With that established, we can observe that because the sequence $\left(\frac{n+1}{2}\right)$ tends to infinity, the sequence (s_n) must also tend to infinity. There are some details to sort out, because (s_{2^n}) is only a subsequence of (s_n), but this is the gist of the argument and you will probably see the details in your Analysis course.

The divergence of the harmonic series should serve as a reminder that intuition based on graphs or finite cases should be generalized to infinite cases only with great caution (you might now like to imagine what a graph of s_n against n would look like for the first million terms). It should also help you to appreciate that infinity is *really* big. The terms of the harmonic series are small and they keep on getting smaller, but there are so many of them that they add up to something infinite anyway. Finally, it highlights the fact that apparently similar series might have dramatically different behaviours:

$$\left(\frac{1}{2^n}\right) \to 0 \text{ and } \sum_{n=1}^{\infty} \frac{1}{2^n} \text{ converges,}$$

but

$$\left(\frac{1}{n}\right) \to 0 \text{ and } \sum_{n=1}^{\infty} \frac{1}{n} \text{ diverges.}$$

This should make you wonder what happens for other series. For instance,

$$\left(\frac{1}{n^2}\right) \to 0; \text{ what do you think happens to } \sum_{n=1}^{\infty} \frac{1}{n^2}?$$

On the one hand this series is a bit 'like' $\sum 1/n$, so maybe it diverges. On the other hand, its terms get smaller a lot faster, so maybe it converges. I will leave this as a cliffhanger for your course.

6.6 Tests for convergence

Any course in Analysis will establish convergence or divergence for 'standard' series like those considered so far in this chapter. It will also introduce and prove numerous tests for convergence that can be applied to more complicated-looking series. I will not include a lot of proofs here, but I will include several tests so that I can draw your attention to some relationships between them. Here is one, for instance:

Theorem (shift rule for series):
Suppose $N \in \mathbb{N}$. Then $\sum a_n$ converges if and only if $\sum a_{N+n}$ converges.

Can you see why this is called the *shift rule*? If $N = 10$, say, it just means that $a_1 + a_2 + a_3 + \dots$ converges if and only if $a_{11} + a_{12} + a_{13} + \dots$ converges.

It doesn't mean that they converge to the same number—obviously chopping ten terms off the beginning will make the series add up to something different, but it cannot make a convergent series divergent (or vice versa). Think about why.

Here is another test:

Theorem (comparison test for series):
Suppose that $0 \le a_n \le b_n \; \forall n \in \mathbb{N}$. Then

1. If $\sum b_n$ converges then $\sum a_n$ converges;
2. If $\sum a_n$ diverges then $\sum b_n$ diverges.

This is intuitively natural, so much so that I invoked it twice in Section 6.4 and you probably didn't notice. Where did I do that, exactly? In fact there are a few related comparison tests, including this one:

Theorem (limit comparison test):
Suppose that $a_n, b_n > 0 \; \forall n \in \mathbb{N}$ and $\left(\dfrac{a_n}{b_n} \right) \to l \neq 0$.

Then $\sum a_n$ converges if and only if $\sum b_n$ converges.

The limit comparison test is useful for establishing results about complicated-looking series by noticing that they are in some sense 'like' simpler ones. For instance, let

$$a_n = \frac{n^2 + 6}{3n^3 - 4n}.$$

Then $\sum a_n$ diverges, because $\sum b_n = \sum 1/n$ diverges and

$$\frac{a_n}{b_n} = \frac{n^3 + 6n}{3n^3 - 4n} = \frac{1 + \frac{6}{n^2}}{3 - \frac{4}{n^2}} \to \frac{1}{3} \text{ as } n \to \infty.$$

Make sure you can see how this relates to the statement of the test.

Where the limit comparison test uses ratios of corresponding terms of two different series, the *ratio test* uses ratios of adjacent terms of the same series:

Theorem (ratio test for series):

Suppose that $a_n > 0 \; \forall n \in \mathbb{N}$ and that $(a_{n+1}/a_n) \to l$ as $n \to \infty$. Then:

1. If $l < 1$ then $\sum a_n$ converges.
2. If $l > 1$ (including $l = \infty$) then $\sum a_n$ diverges.

We'll do two things here: apply this test to a specific series and examine a proof. We will apply it to the series below, which was introduced in Section 6.3. What do you think? Does this series converge or diverge?

$$\sum_{n=1}^{\infty} \frac{n^2}{2^n} = \frac{1^2}{2^1} + \frac{2^2}{2^2} + \frac{3^2}{2^3} + \frac{4^2}{2^4} + \frac{5^2}{2^5} + \dots$$

To find out using the ratio test, we need to consider a_{n+1}/a_n. The form of the series terms means that things cancel when we do this:

$$\frac{a_{n+1}}{a_n} = \frac{(n+1)^2}{2^{n+1}} \cdot \frac{2^n}{n^2} = \frac{1}{2}\left(\frac{n+1}{n}\right)^2 = \frac{1}{2}\left(1 + \frac{1}{n}\right)^2.$$

Now as $n \to \infty$, $\left(1 + \frac{1}{n}\right)^2 \to 1$ so $\frac{1}{2}\left(1 + \frac{1}{n}\right)^2 \to \frac{1}{2}$. This is a limit $l < 1$ so the ratio test tells us that the series converges. That's all there is to applying the ratio test. But people sometimes find it confusing, I think because it gives information about the series via the limit of the sequence of ratios of its terms, which is obviously a complicated chain of reasoning. Check that you can see what I mean, then try applying the ratio test to find out about the convergence or otherwise of the series

$$\sum_{n=1}^{\infty} \frac{2^n}{n!} = \frac{2^1}{1!} + \frac{2^2}{2!} + \frac{2^3}{3!} + \frac{2^4}{4!} + \frac{2^5}{5!} + \dots.$$

Why does the cancelling work even better in this case? And why is it important to remember that we use a_{n+1}/a_n, not a_n/a_{n+1}?

To understand why the ratio test works, we need a proof. I will state the test again, together with a proof for part 1. I like this proof because it is a nice example of theory building: it uses the definition of convergence for the sequence (a_{n+1}/a_n) (see Sections 5.5 and 5.6), the result about convergence for geometric series (Section 6.4) and the comparison test and the shift rule (this section). It also cleverly constructs a number less than 1 by using the fact that $l < 1$. This diagram will help you to see how:

$$\varepsilon = \tfrac{1}{2}(1 - l)$$

With that in mind, have a go at reading the proof (don't forget the self-explanation training from Section 3.5).

Theorem ratio test for series:

Suppose that $a_n > 0 \; \forall n \in \mathbb{N}$ and that $(a_{n+1}/a_n) \to l$ as $n \to \infty$. Then:

1. If $l < 1$ then $\sum a_n$ converges.
2. If $l > 1$ (including $l = \infty$) then $\sum a_n$ diverges.

Proof of part 1: Suppose $a_n > 0 \; \forall n \in \mathbb{N}$ and $(a_{n+1}/a_n) \to l < 1$.

Then, using $\varepsilon = \tfrac{1}{2}(1 - l)$ in the definition of $(a_{n+1}/a_n) \to l$,

$\exists N \in \mathbb{N}$ s.t. $\forall n > N$,

$$\left| \frac{a_{n+1}}{a_n} - l \right| < \tfrac{1}{2}(1 - l) \quad \Rightarrow \quad \frac{a_{n+1}}{a_n} < l + \tfrac{1}{2}(1 - l) = \tfrac{1}{2}(1 + l) < 1.$$

This means that $\forall n > N$, $a_{n+1} < \tfrac{1}{2}(1 + l)a_n$.

In particular,

$$a_{N+2} < \tfrac{1}{2}(1 + l)a_{N+1}$$

and

$$a_{N+3} < \tfrac{1}{2}(1 + l)a_{N+2} < \left(\tfrac{1}{2}(1 + l) \right)^2 a_{N+1}$$

and, by induction,

$$a_{N+n} \leq \left(\tfrac{1}{2}(1 + l) \right)^{n-1} a_{N+1} \quad \forall n \in \mathbb{N}.$$

Now $\sum \left(\tfrac{1}{2}(1 + l) \right)^{n-1} a_{N+1}$ converges because it is a geometric series with common ratio less than 1.

So $\sum a_{N+n}$ converges by the comparison test for series.

So $\sum a_n$ converges by the shift rule for series.

As usual, I'd advise imagining that you are explaining the proof to someone else. Where, if anywhere, do you get stuck? Make a couple of notes on that and you'll be ready to listen for a lecturer's explanation in your course. If you didn't get stuck, can you adapt the argument to prove part 2? Either way, be ready for lots of practice in applying these tests and more.

6.7 Alternating series

Many of the series we've dealt with have only positive terms, but we did look at one with both positive and negative terms:

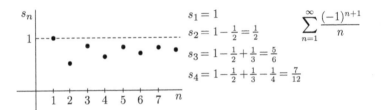

$$s_1 = 1$$
$$s_2 = 1 - \frac{1}{2} = \frac{1}{2}$$
$$s_3 = 1 - \frac{1}{2} + \frac{1}{3} = \frac{5}{6}$$
$$s_4 = 1 - \frac{1}{2} + \frac{1}{3} - \frac{1}{4} = \frac{7}{12}$$

$$\sum_{n=1}^{\infty} \frac{(-1)^{n+1}}{n}$$

This series converges, as should be obvious from the graph. In fact, it converges to $\ln 2$. Does this seem reasonable? To answer, we need an approximate value for $\ln 2$, and this is a situation in which undergraduates reach for their calculators then look a bit embarrassed because they know they should be able to work this out. Here's how I'd do it: $x = \ln 2 \Leftrightarrow e^x = 2$, and e is about 2.7, meaning that x must be a bit less than 1. So that seems okay.

I won't prove that this series converges to $\ln 2$ as that takes a bit of machinery. But it is pretty easy to prove that it does converge, by observing three things:

- The odd terms of (s_n) form a decreasing subsequence $(s_{2n-1}) = s_1, s_3, s_5, \ldots$ that is bounded below and therefore must converge;[3]

---

[3] One of the possible theorems from Section 5.4 was 'Every bounded monotonic sequence is convergent'. This is true, and is discussed further in Section 10.5.

- The even terms of (s_n) form an increasing subsequence $(s_{2n}) = s_2, s_4, s_6, \ldots$ that is bounded above and therefore must converge;
- The terms of both subsequences get arbitrarily close to each other because $(1/n) \to 0$.

The proof below formalizes these observations. Reading it will be good practice in thinking about partial sums. If in doubt about how the algebra works, write out s_{2n+1}, perhaps for a specific value of n, and think about that (e.g. $n = 3$ gives $s_7 = 1 - \frac{1}{2} + \frac{1}{3} - \frac{1}{4} + \frac{1}{5} - \frac{1}{6} + \frac{1}{7}$).

Claim: $\sum \dfrac{(-1)^{n+1}}{n}$ converges.

Proof: Let $s_n = \displaystyle\sum_{i=1}^{n} \dfrac{(-1)^{i+1}}{i}$ as usual.

Then $\forall n \in \mathbb{N}$,

$s_{2n+1} - s_{2n-1} = -\dfrac{1}{2n} + \dfrac{1}{2n+1} < 0$ so (s_{2n-1}) is decreasing, and

$s_{2n+2} - s_{2n} = \dfrac{1}{2n+1} - \dfrac{1}{2n+2} > 0$ so (s_{2n}) is increasing.

So $\forall n \in \mathbb{N}$, $s_2 \leq s_{2n} < s_{2n-1} \leq s_1$.

So (s_{2n-1}) and (s_{2n}) are both monotonic and bounded, so they converge.

Finally, suppose that $\lim_{n\to\infty} s_{2n-1} = s$.

Then $\lim_{n\to\infty} s_{2n} = \lim_{n\to\infty} \left(s_{2n-1} - \dfrac{1}{2n} \right) = s - 0 = s$.

So $(s_n) \to s$ and the series is convergent.

Series like this are known as *alternating series*, for the obvious reason that they alternate between positive and negative terms. Where have we seen a divergent alternating series? Convergent alternating series can be split into two kinds with different behaviours. For some convergent alternating series, the series composed of the absolute values of the terms converges too. For example, both

$$\sum (-1)^n \left(\tfrac{3}{4}\right)^n \text{ and } \sum \left(\tfrac{3}{4}\right)^n \text{ converge.}$$

For other convergent alternating series, the series composed of the absolute values of the terms diverges. For example,

$$\sum \frac{(-1)^n}{n} \text{ converges but } \sum \frac{1}{n} \text{ diverges.}$$

This motivates these two definitions:

Definition: $\sum a_n$ is *absolutely convergent* if and only if $\sum |a_n|$ is convergent.

Definition: $\sum a_n$ is *conditionally convergent* if and only if $\sum a_n$ is convergent but $\sum |a_n|$ is not.

Conditionally convergent series turn out to have a very peculiar property, as described in the next section.

6.8 A really surprising example

Consider the series $\sum a_n = 1 - 1 + \frac{1}{2} - \frac{1}{2} + \frac{1}{3} - \frac{1}{3} + \frac{1}{4} - \frac{1}{4} + \frac{1}{5} - \frac{1}{5} \cdots$.

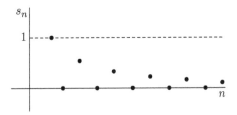

This converges to zero (because the sequence of partial sums (s_n) tends to zero).

Now consider the series $\sum b_n = 1 + \frac{1}{2} - 1 + \frac{1}{3} + \frac{1}{4} - \frac{1}{2} + \frac{1}{5} + \frac{1}{6} - \frac{1}{3} + \cdots$.

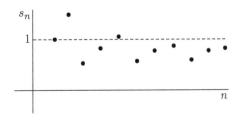

This has the same terms as $\sum a_n$, just in a different order. Make sure you believe that none are missed out. Then notice that if we group the terms in threes, we can rewrite $\sum b_n$ in a simpler way:

$$\sum b_n = \left(1 + \tfrac{1}{2} - 1\right) + \left(\tfrac{1}{3} + \tfrac{1}{4} - \tfrac{1}{2}\right) + \left(\tfrac{1}{5} + \tfrac{1}{6} - \tfrac{1}{3}\right) + \ldots$$
$$= \left(1 - \tfrac{1}{2}\right) \quad + \left(\tfrac{1}{3} - \tfrac{1}{4}\right) \quad + \left(\tfrac{1}{5} - \tfrac{1}{6}\right) \quad + \ldots$$

The result is the alternating series from the Section 6.7, which adds up to ln 2. So, by adding up the terms in a different order, *we get a different sum.*

This is not a trick. It is genuinely possible to add up the terms of this series in different orders and get different totals. If you weren't convinced before that infinite sums behave differently from finite sums, you should be now.

I think this is the weirdest, most counterintuitive result in early Analysis, and this makes series my favourite topic to teach. I get a kick out of it because I like surprises and I especially like understanding how such a counterintuitive result could arise. Some students don't like it so much, because counterintuitive results make them doubt their understanding and become a bit nervous. I will try to explain it in such a way that you can avoid the nerves and share my fascination.

First, don't panic. It is still true that $3 + 5 = 5 + 3$, and indeed order does not matter for any finite sum—adding up a million numbers in any order will give the same total. The peculiar behaviour happens only with infinite series. Indeed, it happens only with conditionally convergent series. Your Analysis lecturer will probably formulate a full algebraic argument explaining why, but it's possible to get the gist of it by understanding an important feature of conditionally convergent series.

In a conditionally convergent series, the terms tend to zero. They must do, because a conditionally convergent series is convergent so it satisfies the null sequence test from Section 6.4. Also, the positive terms alone add up to $+\infty$, and the negative terms alone add up to $-\infty$. You can check this for the series in this section, and you will probably prove it in your course. But this means that conditionally convergent series are truly amazing. Suppose that c is an arbitrary real number. Then, because the positive terms of a conditionally convergent series add up to $+\infty$, we can add them up, keeping them in order, until we get above c. Then, because the negative terms add up to $-\infty$ we can add negative ones, again keeping them in order, until we get below c again. Then we can add positive ones to get above c again, and so on. Keeping the terms in order means we will definitely include them all, so the resulting rearrangement of the series contains the same terms as the original. And the fact that the terms tend to zero means that this process produces a series that converges to c. This means that we can rearrange a conditionally convergent series to make it add up to *any number we want*. Brilliant.

6.9 Power series and functions

The remaining sections in this chapter introduce power series, which pop up all over mathematics. In some courses you will learn to manipulate power series and apply them to practical problems; in others, including Analysis, you'll learn more about the theory. Here I want to make sure that you understand what power series are, how they can be studied using techniques we've already seen, and how they relate to functions.

Definition: A *power series centred at a* is a series of the form

$$\sum_{n=0}^{\infty} c_n(x-a)^n = c_0 + c_1(x-a) + c_2(x-a)^2 + c_3(x-a)^3 + \dots.$$

In particular, a power series centred at 0 is a series of the form

$$\sum_{n=0}^{\infty} c_n x^n = c_0 + c_1 x + c_2 x^2 + c_3 x^3 + \dots.$$

Notice that each term is of the form $c_n x^n$ or $c_n(x-a)^n$, where c_n is a coefficient; the powers of x or $(x-a)$ give power series their name. Notice also that for power series we usually start at $n = 0$ because this allows us to have a constant term, which is desirable because a power series is like an infinite polynomial.

One simple power series is $\sum_{n=0}^{\infty} x^n = 1 + x + x^2 + x^3 + \dots.$

This is just the geometric series with first term 1 and common ratio x. If you've studied some calculus, it's likely that you're also familiar with these power series (what are the coefficients c_n in each case?):

$$1 + x + \frac{x^2}{2!} + \frac{x^3}{3!} + \ldots \qquad 1 - \frac{x^2}{2!} + \frac{x^4}{4!} - \frac{x^6}{6!} + \ldots$$

The first of these is the Maclaurin series of the function $f : \mathbb{R} \to \mathbb{R}$ given by $f(x) = e^x$; the second is the Maclaurin series of the function $g : \mathbb{R} \to \mathbb{R}$ given by $g(x) = \cos x$. But do you really know what it means to say that a series is the Maclaurin series of a function? Many students don't, so we'll sort that out, beginning by clarifying some issues about convergence for such series.

Consider again the series $\displaystyle\sum_{n=0}^{\infty} \frac{x^n}{n!} = 1 + x + \frac{x^2}{2!} + \frac{x^3}{3!} + \ldots.$

This converges for every real number x (you could check using the ratio test), which means that for every x it adds up to a finite number. For $x = 2$ it adds up to one number, for $x = -5$ it adds up to another, and so on. This means that we can treat the series as a function of x, defining

$$f : \mathbb{R} \to \mathbb{R} \text{ by } f(x) = \sum_{n=0}^{\infty} \frac{x^n}{n!}.$$

Because the series is an infinite polynomial in x, this should seem quite natural.

Now consider again the series $\displaystyle\sum_{n=0}^{\infty} x^n = 1 + x + x^2 + x^3 + \ldots.$

This too can be thought of as an infinite polynomial in x, but it does not converge for all values of x. Specifically, it adds up to a finite number only for $x \in (-1, 1)$. So we can treat it as a function of x only for these values, defining

$$f : (-1, 1) \to \mathbb{R} \text{ by } f(x) = \sum_{n=0}^{\infty} x^n.$$

The domain is different because the function is defined only for $x \in (-1, 1)$.

That should clarify what it means to treat a power series as a function, but it doesn't explain how such a function might relate to a familiar one like $g : \mathbb{R} \to \mathbb{R}$ given by $g(x) = \cos x$. We can explain that by thinking about partial sums.

For the series $1 - \dfrac{x^2}{2!} + \dfrac{x^4}{4!} - \dfrac{x^6}{6!} + \ldots$, the first few (distinct) partial sums are

$$1, \quad 1 - \dfrac{x^2}{2!}, \quad 1 - \dfrac{x^2}{2!} + \dfrac{x^4}{4!}, \quad 1 - \dfrac{x^2}{2!} + \dfrac{x^4}{4!} - \dfrac{x^6}{6!}.$$

Each of these can also be thought of as a function of x. So we can plot a few on the same graph as g. Which graph is which?

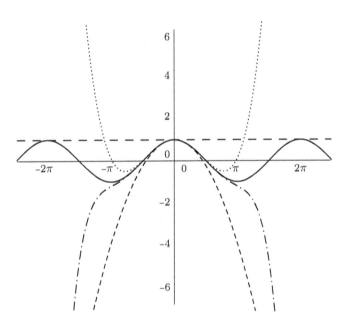

Notice that partial sums with more terms provide better approximations to the function: they 'match' better for more of the curve. If you have access to a graphing calculator or a computer algebra system you might like to plot some graphs for higher powers of n. There are some instructions on one way do this at the end of the chapter, and you might want to flick forward now and look at the graphs in Section 8.8, where we will return to this topic in the context of work on Taylor's theorem.

6.10 Radius of convergence

We have established that some power series converge for every $x \in \mathbb{R}$, and some don't. A question addressed in Analysis is, how can we tell for which x a power series converges? To get some intuition for this it will help to consider an illustrative example, addressed via this extended version of the ratio test:

Theorem (ratio test for series):
Suppose that $|a_{n+1}/a_n| \to l$ as $n \to \infty$. Then:

1. If $l < 1$ then $\sum a_n$ converges.
2. If $l > 1$ (including $l = \infty$) then $\sum a_n$ diverges.

Applying this to the power series $\displaystyle\sum_{n=0}^{\infty} \frac{(x-3)^n}{2n}$, we get

$$\left| \frac{a_{n+1}}{a_n} \right| = \left| \frac{(x-3)^{n+1}}{2(n+1)} \cdot \frac{2n}{(x-3)^n} \right| = \left| (x-3)\left(\frac{n}{n+1}\right) \right|,$$

which tends to $|x-3|$ as $n \to \infty$ (why?). So, by the ratio test, the series converges if $|x-3| < 1$ and diverges if $|x-3| > 1$, meaning that it converges if $2 < x < 4$ and diverges if $x < 2$ or $x > 4$.

The same kind of thing happens for other power series—probably your course will involve numerous examples. Formulating the result in general gives this theorem:

Theorem:
For a power series $\displaystyle\sum_{n=0}^{\infty} c_n(x-a)^n$, exactly one of these is true:

1. The power series converges $\forall x \in \mathbb{R}$.
2. The power series converges only if $x = a$.
3. $\exists R > 0$ such that the power series converges if $|x-a| < R$ and diverges if $|x-a| > R$.

The number R is called the *radius of convergence* of the power series.

Smart students ask two questions at this point. The first is, what if $|x - a| = R$? It turns out that the ratio test does not give decisive information in that case. To get insight into why, think about what happens to the power series we just looked at if $x = 2$ and if $x = 4$.

The second question is, why is this called the *radius* of convergence, when it doesn't involve any circles? The answer to this is very pleasing: it does involve circles, they are just hidden. Everything in this section also works if we take x to be a complex number. In the complex plane, $|x - a| < R$ defines a circle of radius R centred at a—when working with the real numbers, we are just looking at the real-line bit of it:

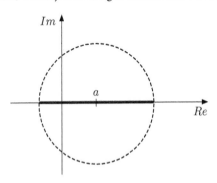

6.11 Taylor series

Many readers will be familiar with the Maclaurin series from Section 6.9, and many will know that we can find the *Taylor series* for a general function f about a point a using this formula (where the notation $f^{(n)}(a)$ means the nth derivative of f at a, and is to be distinguished from $f^n(a)$, which means $f(a)$ raised to the power n):

$$f(x) = f(a) + f'(a)(x - a) + \frac{f''(a)}{2!}(x - a)^2 + \ldots + \frac{f^{(n)}(a)}{n!}(x - a)^n + \ldots$$

To derive the formula, suppose that we can express f as a power series, i.e. by writing

$$f(x) = \sum_{x=0}^{\infty} c_n(x - a)^n = c_0 + c_1(x - a) + c_2(x - a)^2 + c_3(x - a)^3 + \ldots.$$

We need to find the coefficients, and one can be identified immediately: setting $x = a$ gives $f(a) = c_0$, so we have found c_0.

We can find the other coefficients with some judicious differentiation and substitution. Differentiating both sides gives

$$f'(x) = c_1 + 2c_2(x - a) + 3c_3(x - a)^2 + 4c_4(x - a)^3 + \ldots$$

and setting $x = a$ gives $f'(a) = c_1$, so we have found c_1.

Differentiating again gives

$$f''(x) = 2c_2 + 3.2c_3(x - a) + 4.3c_4(x - a)^2 + 5.4c_5(x - a)^3 + \ldots$$

and setting $x = a$ again gives $f''(a) = 2c_2$, so $c_2 = \frac{1}{2}f''(a)$.

Get the idea? It's worth doing one more:

$$f^{(3)}(x) = 3.2c_3 + 4.3.2c_4(x - a) + 5.4.3c_5(x - a)^2 + 6.5.4c_6(x - a)^3 + \ldots$$

and setting $x = a$ gives $f^{(3)}(a) = 3.2c_3$, so $c_3 = \dfrac{f^{(3)}(a)}{3.2}$.

I didn't multiply out the numbers because the structure is easier to see if we don't. Try a couple more steps and you will see that this leads to

$$c_n = \frac{f^{(n)}(a)}{n.(n - 1). \ldots .3.2} = \frac{f^{(n)}(a)}{n!},$$

meaning that the whole series must be the Taylor series

$$f(x) = f(a) + f'(a)(x - a) + \frac{f''(a)}{2!}(x - a)^2 + \ldots + \frac{f^{(n)}(a)}{n!}(x - a)^n + \ldots.$$

Now, that's a nice derivation, and readers who have done a lot of calculus might have seen it before. But in Analysis we do more than just differentiation and algebra—we think about the conditions under which an argument is valid. This derivation shows that *if* a function is equal to a power series about the point a, then that power series must be the Taylor series. But this doesn't tell us under what conditions the 'if' applies. We've looked at a couple of cases (you might know some more) in which the full Taylor series is exactly equal to the function for all values of x. But we've also looked at a function for which that isn't the case. If you go through the above process of differentiation and substitution for the function given by $f(x) = 1/(1 - x)$ about the point $a = 0$, you will end up with

$$\frac{1}{1 - x} = 1 + x + x^2 + x^3 + \dots.$$

But we know that this equality holds only for $x \in (-1, 1)$. The function $f(x) = 1/(1 - x)$ is defined unproblematically for lots of other values of x too, but it is not equal to this power series for those values. There also exist functions that are not equal to their Taylor series anywhere except at $x = a$. These are beyond the scope of this book, but these illustrations should be enough to make you aware that there is a lot to learn here.

6.12 Looking ahead

As in the previous chapter, the preceding sections are just a taster of the material you will study in Analysis. A course that covers series will involve fully formal work on the convergence or otherwise of 'standard' series like

$$\sum x^n \text{ and } \sum \frac{1}{n^\alpha},$$

and proofs of many of the results referred to in this chapter, such as the comparison tests and results on absolute and conditional convergence. You might also study the *integral test*, which provides a link between series and areas under graphs. It is stated here with an accompanying diagram showing the specific case $f(x) = 1/x$—inspecting this and thinking about why it is true (what are the areas of the boxes?) will provide some early experience with ideas that are used to study integrability in Chapter 9.

Theorem (integral test):

Suppose that $f : [1, \infty) \to \mathbb{R}$ is positive and decreasing.

Then $\displaystyle\sum_{n=1}^{\infty} f(n)$ and the sequence $\left(\displaystyle\int_1^n f(x)\mathrm{d}x \right)$ either both converge or both tend to infinity.

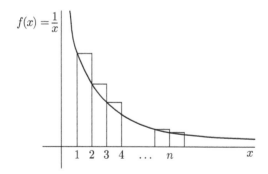

Such rules and tests, when combined with knowledge about a few standard series, can be used to establish the convergence or otherwise of many more. And you will get lots of practice in working out which tests to apply in order to establish convergence for general series and to find radii of convergence for power series.

Ideas about series are also used in more advanced and more applied courses. For instance, Fourier analysis develops the idea of approximating functions using infinite series, in particular combining cosine and sine functions to approximate other periodic functions. Complex analysis generalizes many results to series and power series for which the terms are allowed to be complex numbers, and develops some profound results linking series and functions. This book sticks to real numbers, but we will come back to that link in Chapter 8. In the meantime, here is the promised description of how to generate polynomial approximations to the cosine function (using geogebra, which you can download free from <http://www.geogebra.org>).

1. In the bottom input line, type $f(x) = \cos(x)$ and hit return.
2. Click the second-from-the-right top button to add a slider. Click on the screen where you want to put it. Call the number 'n' and make it run from 0 to 100 in increments of 1, then click 'apply'.
3. In the bottom line, type Taylorpolynomial[f,0,n] and hit return. This produces the graph of the nth partial sum of the power series approximation to $f(x) = \cos(x)$ about the point $a = 0$.
4. Click the left-hand top button to get the pointer, and use it to change n with the slider. This is fun, so don't forget to think about what you're looking at.
5. If you want to zoom out, click the right-hand top button to find that option. Click the magnifying-glass cursor on the drawing pad to use it.
6. Of course, you can mess around with all the inputs to explore other functions, points and partial sums.

CHAPTER 7

Continuity

This chapter begins by discussing intuitive conceptions of continuity and introducing functions that differ from those typically encountered in earlier mathematics. It explains the definition of continuity and demonstrates how it can be used to prove that a function is continuous at a point and to prove more general theorems about continuous functions. Finally, it relates continuity to limits and to proofs involving discontinuities.

7.1 What is continuity?

Most people begin an Analysis course with some useful intuitive knowledge about continuity, but students often need to adjust and develop their thinking in order to appreciate the sophisticated conceptualization used by modern mathematicians.

One common intuitive idea is that a function is continuous 'if you can draw it without taking your pen off the page'. This is not bad as a first approximation—for many simple functions it will lead to correct conclusions. But it is factually limited in a fairly trivial sense, and operationally limited in a more serious way. It is factually limited because a graph can show only a finite part of a function from the reals to the reals; we often draw the bit around the origin $(0, 0)$. So what people really mean is that they believe the whole graph would be 'in one connected piece'. The more serious problem is that although the drawing description sounds practical because it is about a physical action, it cannot be used in mathematical reasoning. For instance, one theorem of Analysis is that if both f and g are continuous functions, then $f + g$ is continuous. Probably you find this believable (though you should, as ever, think beyond your first intuitive response—what stops you coming up with weird examples for which it is not true?). But how could we prove this from the idea that f and g have graphs that 'can be drawn without taking your pen off the page'? We can't, because this is not a manipulable symbolic definition; we can't combine it with function addition. So we need something more mathematical and precise.

A sensible first step toward the mathematical definition is to stop thinking about continuity as a property of a whole function, and start thinking of it as a property that a function might have at a point. You almost certainly do this already: most people, for instance, would agree that this piecewise-defined function is not continuous at $x = 1$, but is continuous everywhere else:

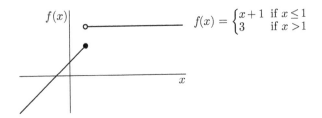

$$f(x) = \begin{cases} x+1 & \text{if } x \leq 1 \\ 3 & \text{if } x > 1 \end{cases}$$

Mathematicians take this as a place to start: they define what it means for a function to be continuous at a point, and only then talk about functions that are continuous everywhere.

In Sections 7.4 and 7.5 I will build up to the definition of continuity through an informal discussion, then take the definition as a starting point and explain what it means. I did the same for the definition of sequence convergence in Chapter 5 and, as in that chapter, you might prefer to read these sections in reverse order. If you have read Chapter 5, you will recognize also that the convergence and continuity definitions have closely related structures. This is handy—although Analysis involves logically complex definitions, they are similar enough that once you get the hang of working with one, others are easier. This occurs because of the close relationship between convergence and continuity, which is discussed in Section 7.6, along with information about common variants of the continuity definition.

Another closely related definition appears in Section 7.9, which discusses limits. It's quite likely that your lecturer or textbook will start with limits then proceed to continuity, but I have chosen to do it the other way around because for me the idea of continuity is more intuitively natural. Before looking at any definitions, however, I will provide some information about function notation and introduce some functions that you might not have seen before.

7.2 Function examples and specifications

Before studying Analysis, most mathematics students have learned about a bunch of standard functions. These usually include:

- quadratic functions like $f(x) = x^2 - 3x + 10$;
- cubic functions like $f(x) = -6x^3 + 5x^2 - 3x + 10$;
- higher-order polynomial functions of the form
$$f(x) = a_0 + a_1x + a_2x^2 + \ldots + a_nx^n;$$
- rational functions like $f(x) = \dfrac{2x^2 - 3x}{x^2 - 5x + 6}$;
- exponential functions like $f(x) = e^x$ or $f(x) = 2^x$;
- logarithmic functions like $f(x) = \ln x$ or $f(x) = \log_{10} x$;
- trigonometric functions like $f(x) = \sin x$;
- inverse trigonometric functions like $f(x) = \tan^{-1} x$
(often written $f(x) = \arctan x$).

Most students can operate with these in a variety of ways, although some are not very quick or reliable at remembering derivatives and integrals or at identifying or sketching graphs. If you suspect that might apply to you, I suggest getting hold of a textbook and doing some practice—you could rely forever on formula booklets and tables, but your mathematical life will be easier if the basics don't slow you down. Here I will discuss function representations in advanced mathematics.

In advanced mathematics in general and in Analysis in particular, people take care over function domains. This is part of a general move toward proper specification of mathematical objects. For instance, Chapter 2 included phrases like this:

Let $f : [0, 10] \to \mathbb{R}$,

which is read aloud as

'Let f be a function from the closed interval $[0, 10]$ to the reals.'

The domain $[0, 10]$ appears explicitly in the specification, after the colon and before the arrow. Section 2.4 gave one reason why: functions might have different properties on different domains. For instance,

$f : [0, 10] \to \mathbb{R}$ given by $f(x) = x^2$ is bounded above, but
$f : \mathbb{R} \to \mathbb{R}$ given by $f(x) = x^2$ is not.

This attention to domains means that if you're asked to give examples, you should make sure they are properly defined. For instance, when I ask Analysis students for an example of a function $f : \mathbb{R} \to \mathbb{R}$ that is not continuous at zero, the most common answer is $f(x) = 1/x$. Clearly this function is not continuous at zero. But unfortunately it is not a function from $\mathbb{R}$ to $\mathbb{R}$. Specifically, it is not defined at zero. It's fine to adapt it by specifying a value at zero, perhaps like this:

$$f(x) = \begin{cases} 1/x & \text{if } x \neq 0 \\ 0 & \text{if } x = 0 \end{cases}.$$

This function is not continuous at zero and it is defined everywhere. Not all lecturers will point out this kind of thing, but they'll think you're a bit mathematically naive if you're not alert to it.

It gets more complicated, too. Students often—quite reasonably—think that we will be careful about the image or range of a function in the same way, so that this must be wrong:

$f : \mathbb{R} \to \mathbb{R}$ given by $f(x) = x^2$.

They think that because the function values are never negative we should write this instead:

$f : \mathbb{R} \to [0, \infty)$ given by $f(x) = x^2$.

In fact, the former is fine. Mathematicians make a distinction between the *codomain*—the set to which the function maps the domain values—and the *image*—the set of values that actually get 'hit' by the function (you might be more accustomed to the term 'range', but this is sometimes used ambiguously so I will avoid it). In Analysis, you will not often be asked about images, and people usually just put '$\hookrightarrow \mathbb{R}$' for everything and don't worry any more about it.

By the way, did you wonder about the notation $[0, \infty)$? Whenever we want an interval to extend to infinity we use a round bracket, because ∞ is not a number so it cannot be said to be 'in' an interval, let alone to be 'the largest number' in an interval.

There do occur instances in which a codomain might be specified more tightly, but again the same idea applies. You might, for instance, see a theorem like this:

Theorem: Suppose that $f : [0, 1] \to [0, 1]$ is continuous.
Then $\exists\, c \in [0, 1]$ such that $f(c) = c$.

Students often interpret the premise of this theorem to mean that the function has to hit all the values in $[0, 1]$, so that it would apply to the function on the left but not to the two functions to the right.

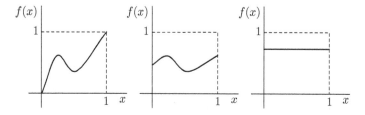

That's not correct. The $[0, 1]$ after the arrow refers to a codomain, not an image, so the premise just means that every $f(x)$ must be in $[0, 1]$, which is true for all three functions. In mathematical terms, f does not have to be *surjective*.

Incidentally, the conclusion of the theorem is about the existence of a *fixed point*—can you see why the point c would be called that? And why must the statement be true?

7.3 More interesting function examples

Analysis involves a lot of theorems about general functions, which are often denoted simply as f or g. If you see this notation and think only of simple, familiar functions, your understanding of the theorems will be limited. Theorems apply to every function for which the premises hold, including weird ones I could make up like this:

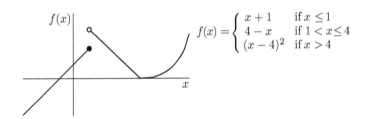

$$f(x) = \begin{cases} x + 1 & \text{if } x \le 1 \\ 4 - x & \text{if } 1 < x \le 4 \\ (x - 4)^2 & \text{if } x > 4 \end{cases}$$

This function is defined piecewise but it is a single function: for every $x \in \mathbb{R}$, it assigns a unique value $f(x)$. We could even construct a function by assigning a random number to each $x \in \mathbb{R}$. Of course, such functions would be hard to work with, and you won't see them very often. But a theorem doesn't care about the specifics of a function provided that its premises apply, and a function does not have to be specified by a single, simple formula, or even by a formula at all.

That said, it is often useful to think about specific functions. It's just that functions that are useful in this context go beyond the familiar ones in the list on page 121. Specifically, those in the list might not be defined everywhere, but where they are defined they are continuous (check to make sure that you believe this). Other functions do not satisfy this property. This one, for instance, is discontinuous at infinitely many points ('$\mathbb{Z}$' denotes the set of all integers—all the whole numbers):

$$f(x) = \begin{cases} 2 & \text{if } x \in \mathbb{Z} \\ 1 & \text{if } x \notin \mathbb{Z} \end{cases}$$

And what about this function?

$$f(x) = \begin{cases} 1 & \text{if } x \in \mathbb{Q} \\ 0 & \text{if } x \notin \mathbb{Q} \end{cases}.$$

This is defined differently depending on whether or not x is *rational*; that is, on whether or not it can be expressed in the form p/q, where $p, q \in \mathbb{Z}$ and $q \neq 0$ (see Section 10.2). The rational and irrational numbers are distributed in a complicated way on the number line—whatever rational number we pick, there are irrationals as close as you like, and vice versa. So this function is not continuous anywhere and it is impossible to draw its graph in a realistic way. However, it can be represented roughly by drawing dotted lines along $f(x) = 0$ and $f(x) = 1$. This is satisfactory provided we remember that it isn't really accurate.

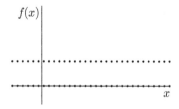

How might we sketch a dodgy-but-potentially-helpful graph for this function?

$$f(x) = \begin{cases} x & \text{if } x \in \mathbb{Q} \\ 0 & \text{if } x \notin \mathbb{Q} \end{cases}.$$

Mathematicians use examples like these in Analysis courses because they can help to clarify the meanings of concepts like continuity and differentiability. But simpler functions can be useful for this purpose too. Consider, for instance, the functions below. Would you say that each one is continuous at zero? Differentiable at zero?

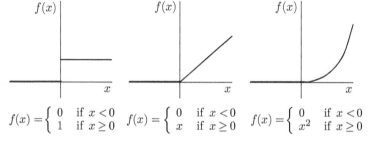

$$f(x) = \begin{cases} 0 & \text{if } x < 0 \\ 1 & \text{if } x \geq 0 \end{cases} \quad f(x) = \begin{cases} 0 & \text{if } x < 0 \\ x & \text{if } x \geq 0 \end{cases} \quad f(x) = \begin{cases} 0 & \text{if } x < 0 \\ x^2 & \text{if } x \geq 0 \end{cases}$$

If you find yourself hesitating, you're not alone—faced with these examples, many students realize that perhaps they don't have a solid grip on those concepts. Differentiability is the subject of the next chapter; this one goes on to explore the definition of continuity.

7.4 Continuity: intuition first

This section starts with informal descriptions of continuity and works up to the definition (if you have read Section 5.5 you will recognize a lot of the reasoning). If you would prefer to start with the definition and read an explanation of how to understand it, you might like to read Section 7.5 before this one.

To begin, remember that mathematicians define continuity at a point. Suppose that f is continuous at a, where f takes the value $f(a)$. How might we capture that idea? Many people say something like 'as x gets closer to a, $f(x)$ gets closer to $f(a)$'. This is a sensible start, but it is not enough, which you can see by looking at these images showing one function that is continuous at a and one that is not. In both cases it is true that as x gets closer to a, $f(x)$ gets closer to $f(a)$. It's just that in the case on the right it doesn't get close enough. That's a shame.

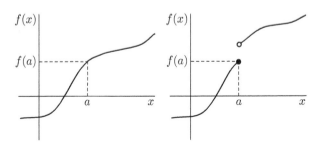

To improve on this, we'll formalize the informal but mathematically appropriate description below.

Informal description: A function f is continuous at a if and only if, by making x close enough to a, we can make $f(x)$ as close as we like to $f(a)$.

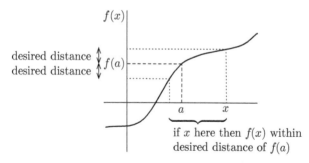

if x here then $f(x)$ within
desired distance of $f(a)$

Notice that if the desired distance were smaller, we might have to make x closer to a. Also, this description excludes the non-continuous function: on the right-hand side of a, even for x really close to a, the vertical 'gap' means that $f(x)$ is never 'close enough' to $f(a)$:

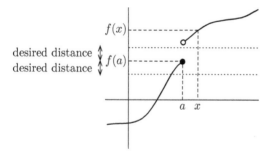

To convert the informal description into a formal definition, we need to get an algebraic handle on the idea of 'close'. Suppose we want $f(x)$ within distance ε of $f(a)$ (as noted in Section 5.5, 'ε' is the Greek letter *epsilon* and is not to be confused with the letter 'e').

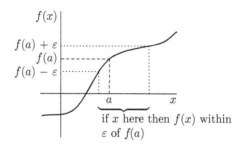

if x here then $f(x)$ within
ε of $f(a)$

So we want $f(a) - \varepsilon < f(x) < f(a) + \varepsilon$, which can be written in the more abbreviated form $|f(x) - f(a)| < \varepsilon$ because

$$|f(x) - f(a)| < \varepsilon \Leftrightarrow -\varepsilon < f(x) - f(a) < \varepsilon$$
$$\Leftrightarrow f(a) - \varepsilon < f(x) < f(a) + \varepsilon.$$

In this context I always read $|f(x) - f(a)| < \varepsilon$ as 'the distance between $f(x)$ and $f(a)$ is less than ε'.

Now, the whole description says '... by making x close enough to a, we can make $f(x)$ as close as we like to $f(a)$.' To capture the notion of 'close enough', mathematicians write the following ('δ' is the Greek letter *delta*).

$$\exists \delta > 0 \text{ such that if } |x - a| < \delta \text{ then } |f(x) - f(a)| < \varepsilon.$$

Don't forget to read the symbolic sentence aloud and think about how each part relates to the diagram. I think about it like this:

$\exists \delta > 0$	such that	if $|x - a| < \delta$,	then $|f(x) - f(a)| < \varepsilon$.
there is a	such	if the distance between x	then the distance between $f(x)$
distance delta	that	and a is less than delta	and $f(a)$ is less than epsilon.

The diagram below has two obvious candidates for the distance δ, a smaller one and a bigger one. Which do we want? The answer is the smaller one; using the bigger distance would include some x-values to the left of a for which $|f(x) - f(a)| \not< \varepsilon$.

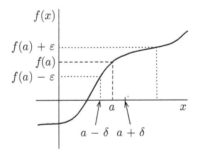

All this is only for one value of ε, however. If we imagine a small ε, this captures the notion that we can make $f(x)$ close to $f(a)$ by making x close to *a*. But it doesn't capture the idea that we can make it *as close as we like*. To do that, it helps to imagine ε getting smaller; we need it to be true that for every $\varepsilon > 0$ we can force $f(x)$ to be within distance ε of $f(a)$ by making x close enough to *a*. This leads to the whole definition:

Definition: A function $f : \mathbb{R} \to \mathbb{R}$ is *continuous at a* $\in \mathbb{R}$ if and only if

$\forall \varepsilon > 0 \exists \delta > 0$ such that if $|x - a| < \delta$ then $|f(x) - f(a)| < \varepsilon$.

If you find it helpful to continue the informal thinking, you could think of it this way:

Definition: A function $f : \mathbb{R} \to \mathbb{R}$ is *continuous at a* $\in \mathbb{R}$ if and only if

$\forall \varepsilon > 0 \quad \exists \delta > 0 \quad$ such that $\quad$ if $|x - a| < \delta, \quad$ then $|f(x) - f(a)| < \varepsilon$.

however small $\quad$ there is a $\quad$ such $\quad$ if the distance between $x \quad$ then the distance between $f(x)$
epsilon is $\quad$ distance delta $\quad$ that $\quad$ and *a* is less than delta $\quad$ and $f(a)$ is less than epsilon.

If you have already begun an Analysis course you might have seen this definition expressed slightly differently, or you might have seen a variant that involves limits or perhaps sequences. I will discuss those briefly in Section 7.6, after an explanation that starts with the definition.

7.5 Continuity: definition first

In this section I will start with the definition of continuity and explain one way to break it down and understand it. If you've read Section 7.4, you will see the same ideas reconstructed in the opposite order, and you might want to use this section to focus on ways of breaking down logical sentences. Here is the definition:

Definition: A function $f : \mathbb{R} \to \mathbb{R}$ is *continuous at a* $\in \mathbb{R}$ if and only if

$\forall \varepsilon > 0 \exists \delta > 0$ such that if $|x - a| < \delta$ then $|f(x) - f(a)| < \varepsilon$.

This definition is about f being continuous at a, where a is treated as a fixed point of interest. The definition also involves a general point x and its associated value $f(x)$. This notation is pretty standard—mathematicians often use letters from the beginning of the alphabet for things that are (at least temporarily) constant and letters from the end for things that are variable. If you have studied Chapter 5, you might like to refer back to the definition of sequence convergence in Section 5.6—its logical structure is very similar. We will break this definition down and understand it in a similar way, starting not at the beginning but at the end.

The last part says $|f(x) - f(a)| < \varepsilon$. This can be read as 'the distance between $f(x)$ and $f(a)$ is less than epsilon'. That's because

$$|f(x) - f(a)| < \varepsilon \Leftrightarrow -\varepsilon < f(x) - f(a) < \varepsilon$$
$$\Leftrightarrow f(a) - \varepsilon < f(x) < f(a) + \varepsilon,$$

so $f(x)$ is between $f(a) - \varepsilon$ and $f(a) + \varepsilon$. We can label such a restriction on the vertical axis of a graph, and adding some dotted lines allows us to see for which values of x it is true that $|f(x) - f(a)| < \varepsilon$. This captures something important about continuity; for the non-continuous function below there are no such x-values to the right of a.

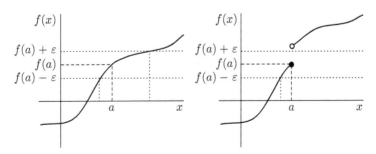

Working backwards through the definition, we get

$$\text{if } |x - a| < \delta \text{ then } |f(x) - f(a)| < \varepsilon.$$

In words, if the distance between x and a is less than delta, then the distance between $f(x)$ and $f(a)$ is less than epsilon. For the function that is continuous at a, we can identify an appropriate δ as in the following diagram. Notice that there are two obvious candidate distances and I chose the smaller one. Why? If you are not sure, think about what would happen to the left of a if we chose the larger one instead.

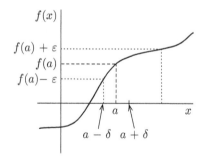

Taking another step back gives

$$\exists\, \delta > 0 \text{ such that if } |x - a| < \delta \text{ then } |f(x) - f(a)| < \varepsilon.$$

To see the significance of this, notice there might not exist an appropriate δ for the function that is not continuous at a—for the labelled value of ε in the next diagram, there is no appropriate δ (to the right of a, even if x is really close to a, it is not true that $|f(x) - f(a)| < \varepsilon$). This is good—the definition classifies these functions as expected.

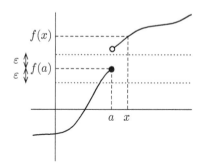

What about the remaining bit?

$$\forall \varepsilon > 0 \; \exists\, \delta > 0 \text{ such that if } |x - a| < \delta \text{ then } |f(x) - f(a)| < \varepsilon.$$

This says that *for all epsilon greater than 0*, the stuff we've already looked at is true. For the continuous function, we could imagine allowing ε to vary: for smaller values of ε, we might need smaller values of δ, but such values would still exist.

Overall, I like to capture the relationship between the informal and formal ideas by thinking of the definition, the diagram, and this informal interpretation:

Definition: A function $f : \mathbb{R} \to \mathbb{R}$ is *continuous at* $a \in \mathbb{R}$ if and only if

$$\forall \varepsilon > 0 \quad \exists \delta > 0 \ \text{such that} \ \text{if } |x - a| < \delta, \quad \text{then } |f(x) - f(a)| < \varepsilon.$$

however small   there is a   such   if the distance between x then the distance between $f(x)$

   epsilon is   distance delta   that   and a is less than delta   and $f(a)$ is less than epsilon.

7.6 Variants of the definition

The definition presented above is standard in Analysis, but you might see variations in how it is written. Some of these are just about notation or style. Others are more about substance and subtleties. I will cover some variants here, but bear in mind that even if the definition in your course looks a bit different, it almost certainly has the same logical structure—if you can't see this after some careful comparing, ask your lecturer or tutor.

First, you might see variation in the phrasing of the 'if...then' clause; perhaps one of these:

Definition: A function $f : \mathbb{R} \to \mathbb{R}$ is *continuous at* $a \in \mathbb{R}$ if and only if

$$\forall \varepsilon > 0 \, \exists \delta > 0 \text{ such that } |x - a| < \delta \Rightarrow |f(x) - f(a)| < \varepsilon.$$

Definition: A function $f : \mathbb{R} \to \mathbb{R}$ is *continuous at* $a \in \mathbb{R}$ if and only if

$$\forall \varepsilon > 0 \, \exists \delta > 0 \text{ such that } \forall x \in \mathbb{R} \text{ with } |x-a| < \delta, \text{ we have } |f(x)-f(a)| < \varepsilon.$$

Second, as noted in Chapter 1, some mathematicians think it is confusing to learn new concepts and new symbols at the same time, so they avoid the quantifier symbols and write everything out in words, like this:

Definition: A function $f : \mathbb{R} \to \mathbb{R}$ is *continuous at* $a \in \mathbb{R}$ if and only if

for all $\varepsilon > 0$ there exists $\delta > 0$ such that if $|x-a| < \delta$ then $|f(x)-f(a)| < \varepsilon$.

If you prefer the words yourself, that's fine—as long as you get the logic right, no one will care.

Third, some people go the other way and write everything in symbols alone, perhaps bracketing various parts of the sentence to indicate what goes with what:

Definition: A function $f : \mathbb{R} \to \mathbb{R}$ is *continuous at* $a \in \mathbb{R}$ if and only if

$$(\forall \varepsilon > 0)(\exists \delta > 0)(|x - a| < \delta \Rightarrow |f(x) - f(a)| < \varepsilon).$$

Fourth and more substantially, these versions of the definition implicitly assume that f is defined for every $x \in \mathbb{R}$ (they talk about $f(x)$ without any indication that it might not be defined). You might see a version that does not make this assumption, but instead defines f only on a restricted domain:

Definition: A function $f : A \to \mathbb{R}$ is *continuous at* $a \in A$ if and only if

$\forall \varepsilon > 0 \exists \delta > 0$ such that if $x \in A$ and $|x - a| < \delta$ then $|f(x) - f(a)| < \varepsilon$.

This is perhaps a better definition, but it is a bit longer, so I will stick to the simpler version for functions that are defined everywhere.

Fifth, you might see this version, where the notation '$\lim_{x \to a} f(x)$' is read aloud as 'the limit as x tends to a of $f(x)$':

Definition: A function $f : \mathbb{R} \to \mathbb{R}$ is *continuous at* $a \in \mathbb{R}$ if and only if $\lim_{x \to a} f(x)$ exists and is equal to $f(a)$.

Indeed, you might have seen this already if you have studied some calculus. It looks different from the definition I gave but in fact it is not, because the definition of limit is closely related to the definition of continuity, as will be discussed in Section 7.10.

Finally, if your course covers both sequences and continuity, you might see this:

Definition: A function $f : \mathbb{R} \to \mathbb{R}$ is *continuous at* $a \in \mathbb{R}$ if and only if for every sequence (x_n) such that $(x_n) \to a$, $\left(f(x_n)\right) \to f(a)$.

Can you work out why this is a reasonable alternative? Try drawing a diagram of a continuous function, marking a sequence of points $x_1, x_2, x_3, \ldots$ on the x-axis such that $(x_n) \to a$, and thinking about the corresponding values $f(x_1), f(x_2), f(x_3), \ldots$ on the vertical axis.

7.7 Proving that a function is continuous

Once a lecturer has introduced a definition, he or she will usually give an example of an object that satisfies it and prove that this is the case. Here we will prove that the function $f : \mathbb{R} \to \mathbb{R}$ given by $f(x) = 3x$ is continuous at every $a \in \mathbb{R}$. I know you know that this function is continuous—this is a case in which you are supposed to focus less on the result and more on how it can be proved within the theory (the thinking is very similar to that in Section 5.7).

People take different approaches to proof construction, and either you or your lecturer might prefer to go about it purely logically and algebraically. As you know, however, I like diagrams, so I tend to start by sketching one. Here is the definition again, together with a diagram showing the function f, a point a, and an arbitrary-looking value of ε.

Definition: A function $f : \mathbb{R} \to \mathbb{R}$ is *continuous at* $a \in \mathbb{R}$ if and only if

$$\forall \varepsilon > 0 \; \exists \delta > 0 \text{ such that if } |x - a| < \delta \text{ then } |f(x) - f(a)| < \varepsilon.$$

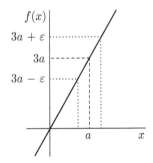

For a given ε, what value of δ will ensure that if $|x - a| < \delta$ then $|f(x) - f(a)| < \varepsilon$? If you can't immediately answer, ask yourself, what if ε is 1? What if ε is $\frac{1}{2}$? And so on. Clearly δ is contingent upon ε: for smaller values of ε, we will need smaller values of δ. In fact, the function 'stretches everything out' by a factor of three, so we need an interval on the x-axis that is three times smaller than the one we want to hit on the $f(x)$-axis. So $\delta = \varepsilon/3$ will work. Having established this, we can start writing a proof by using the structure of the definition as a guide.

We want to prove that the definition holds for every $a \in \mathbb{R}$, so it is sensible to consider an arbitrary one ('arbitrary' in this sense means any a you like, with no assumptions about special properties). For this a, we want to show that $\forall \varepsilon > 0$, something is true. So it makes sense to specify that we are considering an arbitrary $\varepsilon > 0$ too, and start like this:

Claim: $f : \mathbb{R} \to \mathbb{R}$ given by $f(x) = 3x$ is continuous at every $a \in \mathbb{R}$.

Proof: Let $a \in \mathbb{R}$ be arbitrary and let $\varepsilon > 0$ be arbitrary.

For this value of a and this value of ε, we want to show that there exists a value of $\delta > 0$ such that something is true. The easiest way to show that something exists is to produce one, which we can do based on the reasoning above:

Claim: $f : \mathbb{R} \to \mathbb{R}$ given by $f(x) = 3x$ is continuous at every $a \in \mathbb{R}$.

Proof: Let $a \in \mathbb{R}$ be arbitrary and let $\varepsilon > 0$ be arbitrary.

Set $\delta = \varepsilon/3$.

After that, we need to show that if $|x - a| < \delta$ then $|f(x) - f(a)| < \varepsilon$. In this case we can do that by filling in the values for $f(x)$ and $f(a)$ in $|f(x) - f(a)|$ and doing some algebra using the relationship between δ and ε. When reading this, make sure you can see why each equality and inequality is valid.

Claim: $f : \mathbb{R} \to \mathbb{R}$ given by $f(x) = 3x$ is continuous at every $a \in \mathbb{R}$.

Proof: Let $a \in \mathbb{R}$ be arbitrary and let $\varepsilon > 0$ be arbitrary.

Set $\delta = \varepsilon/3$.

Then if $|x - a| < \delta$ we have

$|f(x) - f(a)| = |3x - 3a| = 3|x - a| < 3\delta = 3\varepsilon/3 = \varepsilon.$

The proof is technically done at this point: we have proved that for every $a \in \mathbb{R}$, the definition is satisfied. However, it is polite to write a conclusion line. It would be fine simply to write 'Hence f is continuous at every $a \in \mathbb{R}$,' but you might also like to add an extra line to summarize the argument:

Claim: $f : \mathbb{R} \to \mathbb{R}$ given by $f(x) = 3x$ is continuous at every $a \in \mathbb{R}$.

Proof: Let $a \in \mathbb{R}$ be arbitrary and let $\varepsilon > 0$ be arbitrary.

Set $\delta = \varepsilon/3$.

Then if $|x - a| < \delta$ we have

$|f(x) - f(a)| = |3x - 3a| = 3|x - a| < 3\delta = 3\varepsilon/3 = \varepsilon$.

Hence $\forall a \in \mathbb{R}$ we have shown that

$\forall \varepsilon > 0 \; \exists \delta > 0$ such that if $|x - a| < \delta$ then $|f(x) - f(a)| < \varepsilon$.

So f is continuous at every $a \in \mathbb{R}$ as required.

Obviously you should read a proof like this carefully, following the advice on self-explanation in Section 3.5. You should also think beyond it, asking yourself how it could be modified. For instance, we used $\delta = \varepsilon/3$, but did we need to? Would $\delta = \varepsilon/4$ do instead? How would you modify the proof to deal with $f(x) = 2x + 2$? Or with $f(x) = -3x$? Be careful—I have seen people make this harder than it is by messing up their work with negative numbers and absolute values. How would you modify it to deal with $f(x) = cx$, where c is a constant? Would your modification work for negative values of c? Would it work without further changes for $c = 0$?

One final thing to know is that proofs like this might be presented differently in your course. The proof's structure—as is often the case—directly reflects the structure of the definition. So some people draw diagrams, but it is also common for a lecturer to work algebraically, writing 'Set $\delta = $ ' and leaving a gap, then working out what δ needs to be by generating the chain of inequalities, then going back to fill δ in.

7.8 Combining continuous functions

The proof in the previous section is about a simple linear function, and things get more complicated for other functions. Suppose, for instance, we want to prove that $f : \mathbb{R} \to \mathbb{R}$ given by $f(x) = x^2$ is continuous at every $a \in \mathbb{R}$. The definition could be used to prove this, but what would be the extra complications?

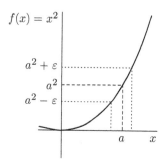

One problem is that the graph is curved, so the appropriate value of δ is contingent not only upon ε but also upon a: the further a is from zero, the smaller δ needs to be. Specifically, a sensible δ would be the smaller of $|\sqrt{a^2 + \varepsilon} - a|$ and $|\sqrt{a^2 - \varepsilon} - a|$ (would this still work for $a < 0$?). Your course might construct a proof based on these observations, but a tidy alternative approach is to prove the product rule instead.

Theorem (product rule for continuous functions):

Suppose that $f : \mathbb{R} \to \mathbb{R}$ and $g : \mathbb{R} \to \mathbb{R}$ are both continuous at $a \in \mathbb{R}$. Then fg is continuous at a.

This theorem can be applied to get the result about $f(x) = x^2$ (how?), and obviously it is a more general theorem.

I won't prove the product rule (a typical proof is similar to that for the product rule for convergent sequences). I will, however, show how the product rule can be used to prove a further theorem, that every function of the form $f_n(x) = x^n$ is continuous everywhere. This is a proof by induction,[1] and it also uses the claim that $f_1(x) = x$ is continuous everywhere—how would you prove that?

---

[1] See any transition-to-proof textbook or Section 6.4 in *How to Study for/as a Mathematics Degree/Major*.

Theorem: For every $n \in \mathbb{N}$, $f : \mathbb{R} \to \mathbb{R}$ given by $f_n(x) = x^n$ is continuous at every $a \in \mathbb{R}$.

Proof: Let $a \in \mathbb{R}$ be arbitrary.

Then $f_1(x) = x$ is continuous at a.

Assume for induction that f_k is continuous at a.

Note that $\forall x \in \mathbb{R}, f_{k+1}(x) = x^{k+1} = x^k x^1 = f_k(x)f_1(x)$.

So $f_{k+1} = f_k f_1$, so f_{k+1} is continuous at a by the product rule.

Hence, by induction, $\forall n \in \mathbb{N}, f_n(x) = x^n$ is continuous at a.

Thus, because a was chosen arbitrarily, we have proved that for every $n \in \mathbb{N}, f : \mathbb{R} \to \mathbb{R}$ given by $f_n(x) = x^n$ is continuous at every $a \in \mathbb{R}$.

Many students understand this proof but wonder why we need to bother with induction, because they think that the product rule proves the theorem directly. But it doesn't. The product rule is about multiplying together exactly two functions—not three, and certainly not n. Theorems do exactly what they say, not more, so induction is necessary here.

In the remainder of this section I want to draw your attention to important differences between some early theorems about continuity. First, here is a theorem and proof that generalizes the result from Section 7.7.

Theorem: Let $c \in \mathbb{R}$. Then $f : \mathbb{R} \to \mathbb{R}$ given by $f(x) = cx$ is continuous at every $a \in \mathbb{R}$.

Proof: Let $a \in \mathbb{R}$ be arbitrary and let $\varepsilon > 0$ be arbitrary.

Set $\delta = \dfrac{\varepsilon}{|c| + 1}$.

Then if $|x - a| < \delta$ we have

$$|f(x) - f(a)| = |cx - ca| = |c||x - a| < |c|\delta = \frac{|c|\varepsilon}{|c| + 1} < \varepsilon.$$

Hence $\forall a \in \mathbb{R}$ we have shown that

$$\forall \varepsilon > 0 \, \exists \, \delta > 0 \text{ s.t. if } |x - a| < \delta \text{ then } |f(x) - f(a)| < \varepsilon.$$

So f is continuous at every $a \in \mathbb{R}$ as required.

That shouldn't seem too bad (though you might like to know that the '+1' in '$|c| + 1$' is just a way to make sure we do not divide by zero). But it is commonly confused with another theorem and proof, especially when people don't think clearly enough about premises and conclusions. Here is the other one.

Theorem (constant multiple rule for continuous functions):
Suppose that $f : \mathbb{R} \to \mathbb{R}$ is continuous at a and $c \in \mathbb{R}$.
Then cf is continuous at a.

> *Proof:* Let $\varepsilon > 0$ be arbitrary.
>
> Then[2] $\exists \delta > 0$ such that if $|x - a| < \delta$ then
> $$|f(x) - f(a)| < \frac{\varepsilon}{|c| + 1}.$$
> Hence $|x - a| < \delta \Rightarrow$
> $$|cf(x) - cf(a)| = |c||f(x) - f(a)| < \frac{|c|\varepsilon}{|c| + 1} < \varepsilon.$$
> So $\forall \varepsilon > 0 \, \exists \delta > 0$ s.t. if $|x - a| < \delta$ then $|cf(x) - cf(a)| < \varepsilon$.
> So cf is continuous at a.

Clearly it is not a good idea to try to learn Analysis by matching symbols, because two theorems and proofs might look very similar but be quite different. Here the first theorem is about the specific function $f : \mathbb{R} \to \mathbb{R}$ given by $f(x) = cx$; the proof shows that this satisfies the definition of continuity. In the second theorem, the fact that f is continuous at a is a premise, and the proof works from this premise to establish that the function cf satisfies the definition of continuity too. The proofs use some similar ideas but their structures are different because of these different premises and conclusions. It might be worth reading them again with this in mind.

7.9 Further continuity theorems

This book does not cover everything that you'll see in an Analysis course. But, as in other chapters, I will list some things that you might see and suggest ways in which you could think about them. First, a theorem that often appears with the product rule:

---

[2] If $\varepsilon > 0$ then $\varepsilon/(|c| + 1) > 0$ so, because f is continuous at a, there must exist a $\delta > 0$ such that if $|x - a| < \delta$ then $|f(x) - f(a)|$ is less than this number $\varepsilon/(|c| + 1)$.

Theorem (sum rule for continuous functions):
Suppose that $f : \mathbb{R} \to \mathbb{R}$ and $g : \mathbb{R} \to \mathbb{R}$ are both continuous at $a \in \mathbb{R}$. Then $f + g$ is continuous at a.

How do you think this might be proved? Try looking at Section 5.10 for inspiration.

Second, a handy lemma:

Lemma: If $f : \mathbb{R} \to \mathbb{R}$ is continuous at $a \in \mathbb{R}$ and $f(a) > 0$ then $\exists \delta > 0$ such that if $|x - a| < \delta$ then $f(x) > 0$.

This lemma is excellent for practising understanding. Do you immediately see what it says and why this is intuitively reasonable (or even obvious)? If not, can you build up a diagram that helps? Sections 2.4 and 2.5 contain advice on how to do that. Once you're convinced that it must be true, think about how it relates to the definition of continuity, and see if you can work out how to prove it.

Third, an obvious theorem:

Intermediate Value Theorem:
Suppose that f is continuous on $[a, b]$ and that y is between $f(a)$ and $f(b)$. Then $\exists c \in (a, b)$ such that $f(c) = y$.

This is worth some thought along the lines of the advice in Section 2.5. Although the theorem is only about values of y between $f(a)$ and $f(b)$, the function might take on values outside the interval:

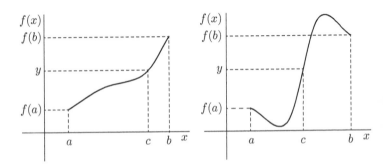

A proof of the Intermediate Value Theorem (IVT) uses some subtle ideas about the properties of the real numbers; I will discuss those in Section 10.5 and will not prove the IVT here. The IVT can be used, however, to prove the fixed point theorem from Section 7.2:

Theorem: Suppose that $f : [0, 1] \rightarrow [0, 1]$ is continuous.
Then $\exists c \in [0, 1]$ such that $f(c) = c$.

Proof: If $f(0) = 0$ or $f(1) = 1$ then we are done.

If not, consider the function $h(x) = f(x) - x$.

h is continuous on $[0, 1]$ by the sum rule.

Now $h(0) > 0$ because $f(0) \in [0, 1]$ but $f(0) \neq 0$,

and $h(1) < 0$ because $f(1) \in [0, 1]$ but $f(1) \neq 1$.

So, by the Intermediate Value Theorem, $\exists c \in (0, 1)$ such that $h(c) = 0$.

But $h(c) = 0 \Rightarrow f(c) = h(c) + c = 0 + c = c$.

So $\exists c \in [0, 1]$ such that $f(c) = c$ as required.

If you would like some graphical intuition to go with the proof, try relating it to the next diagram, in which the function satisfies the premises, the line $y = x$ is shown, and the double-headed arrow represents $h(x)$.

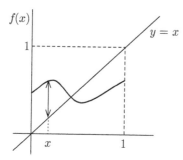

7.10 Limits and discontinuities

Continuity is closely related to limits, and this section will examine how by comparing the relevant definitions. As usual, though, I will start with some diagrams. The two functions shown below are both discontinuous at a. But they are discontinuous in qualitatively different ways. The first does not have a limit as x tends to a: points to the left and right of a have function values that are 'far' from each other. The second function does have a limit as x tends to a: if we imagine approaching a from either side, the function values all get close to the number l, and mathematicians would say that $f(x)$ tends to l as x tends to a. It happens that $l \neq f(a)$, so the function is not continuous at a, but it does have a limit.

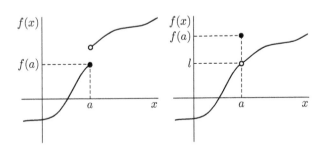

This, to my mind, provides the key insight into the relationship between limits and continuity and explains one variant of the continuity definition:

Definition: A function $f : \mathbb{R} \to \mathbb{R}$ is *continuous at $a \in \mathbb{R}$* if and only if $\lim_{x \to a} f(x)$ exists and is equal to $f(a)$.

In effect, continuity requires the limit to exist and the value $f(a)$ to be 'in the right place'. That's the intuition (or at least a version of it that makes sense to me—look out for other explanations if you are not quite satisfied). The definitions of limit and continuity appear below so that you can compare them.

Definition (limit): $\lim_{x \to a} f(x) = l$ if and only if

$\forall \varepsilon > 0 \exists \delta > 0$ such that if $0 < |x - a| < \delta$ then $|f(x) - l| < \varepsilon$.

Definition (continuity):
A function $f : \mathbb{R} \to \mathbb{R}$ is *continuous at $a \in \mathbb{R}$* if and only if

$\forall \varepsilon > 0 \exists \delta > 0$ such that if $|x - a| < \delta$ then $|f(x) - f(a)| < \varepsilon$.

The limit definition involves a general limit l rather than the value $f(a)$. The only other difference is that for limits we require that something holds if $0 < |x - a| < \delta$, rather than if $|x - a| < \delta$. What effect does this have? Thinking in terms of distances, $0 < |x - a|$ means that the distance between x and a is strictly greater than zero, so $x \neq a$ and the definition says nothing about what happens at the point $x = a$. So a function can have a limit at a without having the 'right' $f(a)$ value. Indeed, it can have a limit even if $f(a)$ is not defined.

The limit version of the continuity definition provides a simple way of thinking about some of the functions cited earlier in this chapter. Of the functions below, for example, only the leftmost is not continuous at zero. The others are both continuous at zero because, although they are defined differently to the left and the right of zero, they each have the same limit from the left and from the right and this limit is equal to $f(0)$.

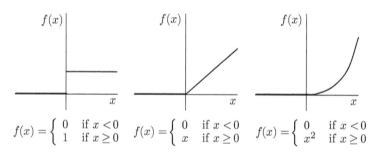

$$f(x) = \begin{cases} 0 & \text{if } x < 0 \\ 1 & \text{if } x \geq 0 \end{cases} \qquad f(x) = \begin{cases} 0 & \text{if } x < 0 \\ x & \text{if } x \geq 0 \end{cases} \qquad f(x) = \begin{cases} 0 & \text{if } x < 0 \\ x^2 & \text{if } x \geq 0 \end{cases}$$

What is required to prove this might depend on the focus of your course. In earlier calculus courses, it is often enough simply to observe that the limits from the left and the right are the same or different. In Analysis, you will probably be expected to prove any such claim from the definition of limit. Because the limit and continuity definitions have similar structures, the information in this chapter about working with the continuity definition should be useful for thinking about how to do that.

An alternative is to prove discontinuity directly by showing that the original definition of continuity is not satisfied. We'll do that for this function, which appeared in Section 7.1:

$$f : \mathbb{R} \to \mathbb{R} \text{ given by } f(x) = \begin{cases} x + 1 & \text{if } x \leq 1 \\ 3 & \text{if } x > 1 \end{cases}.$$

This function is not continuous at 1. Notice that $f(1) = 2$, so to prove this we need to prove that it is *not* true that

$$\forall \varepsilon > 0 \; \exists \delta > 0 \text{ such that if } |x - 1| < \delta \text{ then } |f(x) - 2| < \varepsilon.$$

Because this says 'for all epsilon, there exists delta', we can prove that it is *not* true by showing that there exists an epsilon for which there is no appropriate delta—pause now to make sure you understand this. Here it happens that for $\varepsilon = \frac{1}{2}$, there is no appropriate δ. I will simply present a proof here; the need to negate the definition makes this somewhat logically complex, so take your time in relating each line to the definition and, if you find it helps, to the accompanying diagram.

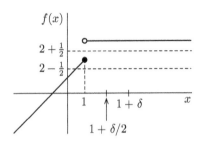

Claim:

$f : \mathbb{R} \to \mathbb{R}$ given by $f(x) = \begin{cases} x + 1 & \text{if } x \leq 1 \\ 3 & \text{if } x > 1 \end{cases}$ is not continuous at 1.

Proof: Note that $f(1) = 2$.

Consider $\varepsilon = \frac{1}{2}$ and let $\delta > 0$ be arbitrary.

Then $x = 1 + \delta/2$ satisfies $|x - 1| < \delta$, but

$$|f(x) - f(1)| = |3 - 2| = 1 > \varepsilon.$$

So for $\varepsilon = \frac{1}{2}$ there does not exist $\delta > 0$ such that

$$\text{if } |x - 1| < \delta \text{ then } |f(x) - f(1)| < \varepsilon.$$

So f is not continuous at 1.

As ever, it is worth thinking about how this proof could be adapted to related cases. What if the $f(1)$ value was 'joined on' to the right-hand branch of the graph rather than the left? What would we need to change for a function with a smaller 'jump'?

For a final note about discontinuities, consider again the two functions below. The function on the left is not continuous anywhere, and you should think about how you would prove this. What about the one on the right? Most people answer that this is not continuous anywhere either. But that answer is based on intuition and not on the definition. In fact, this function is continuous at zero. Imagine $|f(x) - f(0)| < \varepsilon$ on the vertical axis—there would certainly be a δ such that if $|x - 0| < \delta$ then $|f(x) - f(0)| < \varepsilon$, so the definition of continuity is satisfied.

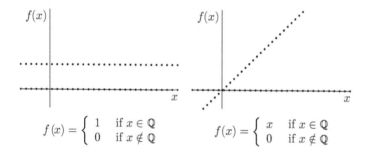

$$f(x) = \begin{cases} 1 & \text{if } x \in \mathbb{Q} \\ 0 & \text{if } x \notin \mathbb{Q} \end{cases} \qquad f(x) = \begin{cases} x & \text{if } x \in \mathbb{Q} \\ 0 & \text{if } x \notin \mathbb{Q} \end{cases}$$

This is a good example of a case in which the mathematical version of a concept corresponds well with most people's intuition for most cases, but classifies some 'boundary' cases differently. This is nothing to worry about—just bear it in mind and remember to work with the definition. And can you construct a function that is continuous at exactly two points? Exactly three? Exactly *n*?

7.11 Looking ahead

Continuity is a substantial topic and, due to the complexity of its central definition, one that can be challenging. It is worth remembering that definitions get easier to handle the more you work with them, so even if the material keeps building up, at some point it will seem easier because you're getting used to it. It is also worth knowing that many Analysis courses involve some work on continuity and then go on to differentiability. Differentiability tends to be easier than continuity, so don't be put off if you get bogged down in some of the more substantial continuity proofs—you'll probably get a fresh start halfway through your course.

A course that involves continuity will probably cover all the material presented here. There will be proofs of theorems for combining continuous functions: the constant multiple, sum, and product rules (and the quotient rule—what do you think that one says?). Often these are grouped together under a heading like 'algebra for continuous functions'. Or similar results might be proved for limits first, then applied directly to continuity using the limit version of the continuity definition. There might be a theorem stating that every polynomial function is continuous everywhere—perhaps you can work out now how to prove that using the sum and product rules rules and proof by induction. There will be a proof and various applications of the Intermediate Value Theorem, and a proof of the Extreme Value Theorem:

Extreme Value Theorem:
Suppose that $f : [a, b] \to \mathbb{R}$ is continuous on $[a, b]$. Then

1. f is bounded on $[a, b]$;
2. $\exists x_1, x_2 \in [a, b]$ such that $\forall x \in [a, b], f(x_1) \leq f(x) \leq f(x_2)$.

This is often stated more briefly as 'a continuous function on a closed interval is bounded and attains its bounds'—can you see why? And it is good for theorem-understanding practice. Draw some diagrams and ask yourself why it has to be true, then try dropping one of the premises—would the conclusion still hold if the function did not have to be continuous? What if it were defined on an open interval (a, b) instead of a closed interval $[a, b]$? And why do you think this is called the Extreme Value Theorem?

In further courses you might learn advanced versions of some of these ideas. A closed interval, for instance, is an example of the more general notion of a *compact set*. In a course on topology, you might learn more about compact sets, and about ways of characterizing continuity using open and closed sets without the restriction that the domain of a function is a subset of $\mathbb{R}$. In further Analysis courses or in work on metric spaces, you might learn about *uniform continuity*, which for real-valued functions is defined like this:

Definition: $f : A \to \mathbb{R}$ is *uniformly continuous* on A if and only if

$\forall \varepsilon > 0 \exists \delta > 0$ such that $\forall x_1, x_2 \in A, |x_1 - x_2| < \delta \Rightarrow |f(x_1) - f(x_2)| < \varepsilon$.

How does this differ from standard continuity? Are there functions that are continuous but not uniformly continuous and vice versa?

In multivariable calculus, you might generalize the concepts of continuity and limits to functions of more than one variable such as $f : \mathbb{R}^2 \to \mathbb{R}$ given by $f(x, y) = x^2 y$. Such functions can be thought of as defining surfaces in three dimensions instead of curves in two dimensions. How do you think continuity and limits might work in such a context? And limits are used to define differentiability, as discussed in the next chapter.

CHAPTER 8

Differentiability

This chapter discusses gradients/slopes and tangent lines, pointing out common misconceptions and explaining how to avoid them. It relates the definition of differentiability to graphical representations, shows how to apply it for simple functions, and demonstrates ways in which a function might fail to be differentiable. It then discusses the Mean Value Theorem and Taylor's Theorem, relating these to graphs and proofs.

8.1 What is differentiability?

This chapter is about *differentiability*, not about differentiation. Probably you've been studying differentiation for a couple of years, and no one doubts your ability to differentiate standard functions or to use formulas to differentiate others that you have never seen before (though you might like to make sure that you can do so quickly and accurately—mathematicians will expect it). In undergraduate mathematics you will have at least one course in which you learn fancier techniques for differentiating more complicated functions. Analysis, however, is not that course.

In Analysis, we are less interested in performing differentiation, and more interested in what it actually means for a function to be differentiable. You might have thought about this a lot, or you might have thought about it a bit but then forgotten about it because it wasn't on an exam, or you might never have thought about it—perhaps you just learned to differentiate by using tables of derivatives. Whichever is the case, one aim of this chapter is to strengthen your knowledge of both intuitive and formal ways of thinking about differentiability, and to make sure that you have a good sense of the links between the two. We will start with an intuitive approach, recalling the ideas raised briefly in Section 2.8.

At a rough first approximation, a function is differentiable at a point if it makes sense to speak of it as having a certain *gradient* or *slope* at that point (people in the UK say 'gradient', people in the US say 'slope'—I'm British so I will use 'gradient' throughout). Equivalently, a function is differentiable at a point if it makes sense to put a tangent line on the graph at that point. Consider, for instance, the linear function $f : \mathbb{R} \rightarrow \mathbb{R}$ given by $f(x) = 3x$. This function has gradient 3 everywhere. You probably haven't thought much about tangent lines for linear functions, because tangents are really only interesting when a graph is curvy rather than straight. But the tangent to this linear graph is the same as the graph itself (to see how this works algebraically, pick a point and follow your usual approach to finding tangent lines). Certainly it makes sense to talk about gradients for functions like this.

Next, consider a nonlinear function, say $f : \mathbb{R} \rightarrow \mathbb{R}$ given by $f(x) = x^2$. Clearly the notion of gradient that applies to functions of the form $y = mx + c$ does not apply to this function.[1] However, an intuitive extension of it does work. The graph is curved, but if we imagine zooming in again and again, what we see looks more and more like a straight line. It never actually is a straight line, but most people would be happy to say that 'in the limit', it is possible to put a sensible tangent line on the graph: we would be satisfied that this line 'matches' the function at a point in the sense that its value is the same and its gradient is the same.

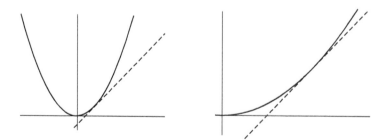

---

[1] In the UK it's $y = mx + c$, in the US it's $y = mx + b$. I have no idea why.

So, for a function like $f(x) = x^2$, the graph is curved but it makes intuitive sense to talk about gradients. We can't talk about the *gradient of the graph*, like we can for $f(x) = 3x$, because the gradient keeps changing. But we can talk about the *gradient at a point*, which is good enough.

However, not all functions are like this. Consider, for instance, the function $f : \mathbb{R} \to \mathbb{R}$ given by $f(x) = |x|$. At most points, this has a sensible gradient: to the left of 0, the gradient is -1; to the right, it is 1. But what about at the point $x = 0$? What happens if we zoom in there? Something different: we can zoom in as much as we like and the graph never looks any straighter. It always has a 'corner', and the corner never gets any less pointy. So it does not make sense to say that the graph has a gradient at zero, and we cannot draw a tangent line that 'matches' the graph.

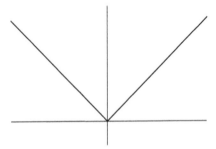

This is what differentiability is about. Informally, a function is differentiable at a point if it makes sense to specify a gradient at that point and therefore to think of a meaningful tangent line at that point. A function is non-differentiable at a point if it has a 'corner' there. This glosses over lots of technical details and more sophisticated considerations, but I want you to think of differentiability as a meaningful concept and this is a decent place to start.

8.2 Some common misconceptions

Before we look at the definition of differentiability, I want to draw your attention to some common misconceptions about derivatives and tangent lines. You might not have any of these misconceptions, but they are tempting ways of thinking and they are common among undergraduates. So we will examine why they are incorrect and get them out of the way.

First, faced with a point on a graph for which there is no meaningful tangent line, people are sometimes tempted to put one in anyway. At a 'corner', they might draw a 'tangent' that is mid-way between the tangents on either side, or even draw several 'tangents', as though imagining the line rotating about the point as it moves from one part of the graph to the other. In relation to accepted mathematical theory, this is wrong. At any point, a graph either has a single meaningful gradient (and therefore a meaningful tangent line) or it doesn't.

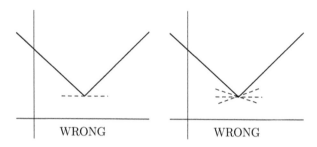

Second, recall that in Section 2.6 I discussed ways in which sketches can be misleading. Something similar happens with gradients. Most people, when they hand-sketch a graph of the function $f : \mathbb{R} \rightarrow \mathbb{R}$ given by $f(x) = \sin x$, draw something like this:[2]

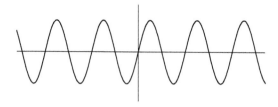

This makes it look like the graph is very steep at, for instance, the point $x = 2\pi$. However, the maximum gradient is actually only 1 (why?). A graph of f with the axes drawn to the same scale actually looks like this:

[2] Actually, a lot of people ignore the bit where $x < 0$. Please don't do that.

There is no problem with the original sketch—we can sketch graphs with whatever axis scales we like—but we should make sure that our interpretation of a graph is informed by the nature of the function, rather than allowing intuition about the function to be swayed by a slightly misleading graph.

Third, some students first meet tangents not in relation to functions but in relation to circles. A tangent to a circle meets the circle at only one point, and does not cross the circle at the point of tangency (it is 'outside' the circle in both directions). This is obvious when looking at a picture:

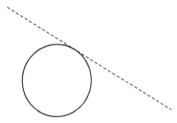

Neither of these things holds for tangent lines to function graphs. It is perfectly possible for a tangent to meet the graph in more than one point, and for it to cross the graph at the point of tangency or at some other point. For instance, consider again the function $f : \mathbb{R} \to \mathbb{R}$ given by $f(x) = \sin x$. Can you identify points at which the tangent crosses the curve at the point of tangency? At some other point? At infinitely many other points? Probably nobody explicitly over-generalizes from circles to function graphs—no student has an internal monologue that goes 'Circles and functions are the same so the tangent must not cross the graph.' But it is important to remember that understanding built up in one area of mathematics might not apply in another.

Fourth, this problem might be exacerbated by confusion over zero derivatives. Most students have no problem with statements like 'The derivative of the function $f(x) = x^2$ is zero at $x = 0$.' However, quite a few get confused when we start talking about the derivative of, say, the function $g : \mathbb{R} \to \mathbb{R}$ given by $g(x) = 5$. I think there are at least three dodgy conceptions bound up in this confusion. Some people are not quite sure that $g(x) = 5$ really is a function. After all, 5 is just a number, and surely a function 'ought to have xs in it'. Again, probably no one explicitly thinks this, but people feel disturbed anyway because the expression seems different from what they were expecting. This confusion is sometimes cleared up with better writing. If we write

the function $g : \mathbb{R} \to \mathbb{R}$ given by $g(x) = 5$ for every $x \in \mathbb{R}$,

that seems a bit better.

Other people think that 'you can't differentiate a number'. That is technically true, but not in the way they mean it. We can't differentiate a number—it is not the right kind of object—but we *can* differentiate a function that everywhere takes on that number as its value, as in the case of $g(x) = 5$. The derivative of g is zero everywhere because g is a constant function so its graph is 'flat'. Indeed, most people who get confused by this would have no problem 'differentiating the 5' in a function like $h : \mathbb{R} \to \mathbb{R}$ given by $h(x) = x^2 + 5$. They do know that 'constants differentiate to zero', they are just (unknowingly and unnecessarily) troubled by a case in which the whole function is constant.

Finally, this might be a case in which the 'zero is nothing' misconception rears its head. Even when looking at a graph of a constant function, people sometimes want to say that it 'doesn't have a derivative', I think because they misinterpret having a zero derivative as not having a derivative. It's easy to see where this comes from: we first learn about numbers by counting objects in the world, and having zero sheep is the same as not having any sheep. But, mathematically, zero is not 'nothing', it is a perfectly good number. If you are willing to say that the derivative of one function is three, you should be willing to say that the derivative of another is zero.

Any or all of these conceptual problems might contribute to an unwillingness to draw a horizontal tangent at the point $x = 0$ for the function $g(x) = x^3$. We also face an extra problem in this case. For $f : \mathbb{R} \to \mathbb{R}$ given by $f(x) = x^2$, for instance, the gradient to the left of $x = 0$ is negative, and the gradient to the right is positive, so it stands to reason that if we travel from left to right we must pass through a point at which the gradient is instantaneously zero (provided we assume that the gradient changes 'smoothly', which is reasonable and which everyone intuitively does). For $g(x) = x^3$, on the other hand, the gradient on the left of $x = 0$ is positive, and the gradient to the right is also positive, so we do not get the same logical leg-up: we are not able to reason that the gradient 'must be zero' at some point.

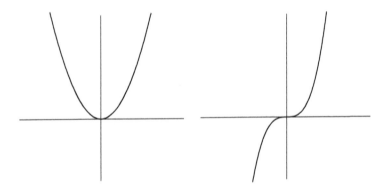

It happens that it *is* zero at some point, but it's easy to see how lazy drawings might lead to people not quite making their derivative zero.

To think about derivatives properly, then, we need to understand them in a fuller, more formal sense. A quick comment first, however. The astute reader will have noticed that every function in this chapter so far is continuous. Having encountered a variety of non-continuous functions in Chapter 7, you might be wondering how the notion of differentiability applies in those cases. In particular, you might have noticed that the notion of a 'corner' makes no sense if a function is not continuous at that point. Have a go now at thinking about how gradients and tangents might apply to non-continuous functions. I will come back to this issue once we have explored the definition of differentiability.

8.3 Differentiability: the definition

Rather than simply presenting the definition of differentiability, I will show how it arises as a natural extension of the notion of a gradient and how it relates to the informal ideas built up in Section 8.2. We will start, as before, with a linear function. You can imagine $f(x) = 3x$ if you like, but I'm going to draw the diagrams without that level of specificity. Informally, the gradient is the answer to the question 'If we travel to the right by one unit, how many units do we go up?' (where down-units count as negative up-units).

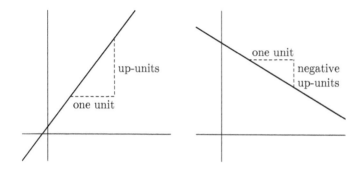

Travelling to the right by exactly one unit is a bit restrictive, however, and we don't have to stick to it. Because of the way ratios work, we get the same number if we take any two points on the graph and ask 'What

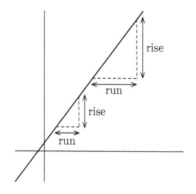

is the ratio of the vertical (upwards) change to the horizontal (rightward) change?' (Americans refer to this economically as 'rise over run').

Adding labels will help when we come to generalize, and there are two common notations in use. The first involves a 'main' point a and a neighbouring point x; the second involves 'main' point x and a neighbouring point $x + h$. Labelling all the appropriate f values, we can write down corresponding expressions for the gradient:

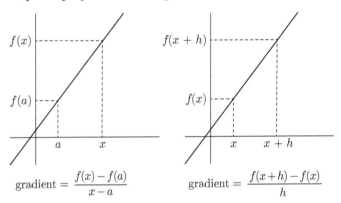

$$\text{gradient} = \frac{f(x) - f(a)}{x - a} \qquad\qquad \text{gradient} = \frac{f(x+h) - f(x)}{h}$$

What would happen if x were to the left of a (or if h were negative)? We ought to get the same answer for the gradient, and we do. I will leave it to you to check—if you are unsure, confirm it algebraically, perhaps with the function $f(x) = 3x$.

To generalize, we can put all the same labels on curved graphs (I'm continuing to show both labelling systems):

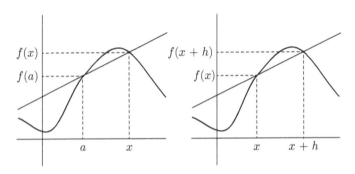

The line joining the points $(a, f(a))$ and $(x, f(x))$ is no longer a tangent to the graph,[3] so we give it a different name, calling it a *secant line*. But the secant line gets 'closer' to being a tangent line if we move x closer to a (I need several diagrams here so I'm using just one labelling system):

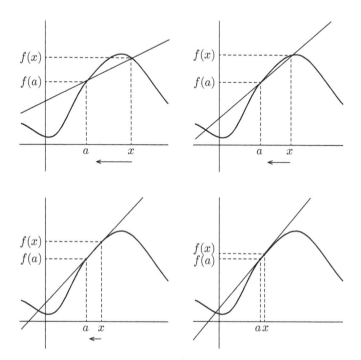

This is the idea used by mathematicians. We imagine the moving point getting closer and closer to the main point, dragging the secant line with it, until, *in the limit*, the secant line 'becomes' the tangent line. To capture this, we write

$$f'(a) = \lim_{x \to a} \frac{f(x) - f(a)}{x - a} \quad \text{or} \quad \frac{df}{dx}\bigg|_a = \lim_{x \to a} \frac{f(x) - f(a)}{x - a}.$$

[3] At least it isn't in general—can you concoct a sketch in which the graph is curved but this process does happen to give a tangent line at some point?

In the first expression, the left-hand side is commonly read aloud as 'f prime of a'; in the second, the left-hand side is read as 'df by dx evaluated at a'. In both, the right-hand side is read as

'the limit as x tends to a of f of x minus f of a over x minus a.'

You might want to check that you could write out the same information using the other labelling system.

We are not done yet, however. These formulas give us derivatives, but in Analysis we are really interested in differentiability. As a result, you will actually see a definition like one of these:

Definition: f is *differentiable at a* if and only if $\lim\limits_{x \to a} \dfrac{f(x) - f(a)}{x - a}$ exists.

Definition: $f'(a) = \lim\limits_{x \to a} \dfrac{f(x) - f(a)}{x - a}$, *provided this limit exists.*

Notice that the first is not about the value of the limit, it is about the *existence* of the limit. The second does define the derivative but, because Analysis is less about calculating values and more about establishing properties like differentiability, the words 'provided this limit exists' are key: the definition is incomplete without them. In either case, if you are asked for a definition of differentiability and you just give the algebraic-looking limit part, you have not written a full definition.

8.4 Applying the definition

Differentiability, then, is about the existence or otherwise of a limit. So we can learn a lot by looking at how such a limit might fail to exist, meaning that a function is not differentiable at a given point. First, though, we will apply the definition to some familiar differentiable functions to confirm that derivatives come out as we expect.

Consider $f : \mathbb{R} \to \mathbb{R}$ given by $f(x) = x^2 + 3x + 1$. Using the formulation involving x and $x + h$, here is how I would write out a proof that this function has derivative $f'(x) = 2x + 3$ (as you read this proof, remember the self-explanation training from Section 3.5).

Claim: If $f(x) = x^2 + 3x + 1$ then $f'(x) = 2x + 3$.

Proof: $\forall x \in \mathbb{R}$ we have

$$\frac{f(x+h) - f(x)}{h} = \frac{(x+h)^2 + 3(x+h) + 1 - x^2 - 3x - 1}{h}$$

$$= \frac{x^2 + 2xh + h^2 + 3x + 3h + 1 - x^2 - 3x - 1}{h}$$

$$= \frac{2xh + h^2 + 3h}{h}$$

$$= 2x + h + 3.$$

So $\forall x \in \mathbb{R}$ we have

$$f'(x) = \lim_{h \to 0} \frac{f(x+h) - f(x)}{h} = \lim_{h \to 0}(2x + h + 3) = 2x + 3.$$

There are several things to notice about the way in which this is written. First, there are two explicit statements that the equations hold for all $x \in \mathbb{R}$. This is good practice because in some cases different values of x give different results, and because it is polite to a reader—better to have too many places where we specify what we are talking about than too few. Second, the proof presents all the algebra for the *difference quotient* $(f(x + h) - f(x))/h$ before discussing its limit. I always advise students to work in this way because people often make errors when they don't. In particular, they tend to put 'lim' in front of the first expression, then forget, so they write things like this:

$$\lim_{h \to 0} \frac{f(x+h) - f(x)}{h} = \frac{(x+h)^2 + 3(x+h) + 1 - x^2 - 3x - 1}{h} = \ldots$$

This equality simply isn't valid: the limit on the left-hand side is not equal to the expression on the right. This error is often exacerbated by the writer remembering at the end and throwing a 'lim' in front of the last expression, but not putting it in for the intermediate ones. I'm not preaching from on high here—this is exactly the kind of error that I make myself, and I find the easiest way to avoid it is to get all the algebra out of the way first, and only then to talk about limits (the result is better in a technical sense, too, because we don't really confirm that the limit exists until the end). Third, there is no particular reason to use the second formulation of the definition here, and trying it with the other labelling system would be a good exercise. Finally, studying the algebra should make it clear that with polynomials, cancelling will always be possible, so that all the derivatives will come out as we expect.

That said, I will now use the other version of the definition to confirm that $g : \mathbb{R} \rightarrow \mathbb{R}$ given by $g(x) = x^3$ has derivative zero at zero, because I want to highlight two useful things.

First, we can work with the point $a = 0$ in isolation:

Claim: If $g(x) = x^3$ then $g'(0) = 0$.

Proof: Note that $\forall x \in \mathbb{R}$ we have $\dfrac{g(x) - g(0)}{x - 0} = \dfrac{x^3 - 0^3}{x - 0} = \dfrac{x^3}{x} = x^2$.

So $g'(0) = \lim\limits_{x \to 0} \dfrac{g(x) - g(0)}{x - 0} = \lim\limits_{x \to 0} x^2 = 0$.

This is neat and quick, and it shows that we do not need to apply a definition to get a derivative for a whole function if we are just interested in the derivative at one point. In this case, the fact that lots of things are zero makes for a quick calculation.

Alternatively, we could work out the general derivative at a then apply it to the case $a = 0$. I will do this now, in the process demonstrating a quick way to do division with polynomials. Notice that

$$\frac{g(x) - g(a)}{x - a} = \frac{x^3 - a^3}{x - a}.$$

It has come to my attention that lots of people have been taught to work with such expressions using rather laborious long division, but my A-level teacher taught me a quicker way to proceed. Here's how it goes.

The question is, by what do we need to multiply $x - a$ to get $x^3 - a^3$? In other words, what is the 'something' in this expression?

$$x^3 - a^3 = (x - a)(\text{something}).$$

We can work that out by thinking about what to put in the brackets. To get the x^3 term, we'll need an x^2:

$$x^3 - a^3 = (x - a)(x^2 \qquad).$$

But now multiplying out the right-hand side will give a term $-ax^2$. We don't want one of those, so we need to make a term $+ax^2$ to cancel it out. We can do that like this:

$$x^3 - a^3 = (x - a)(x^2 + ax \qquad).$$

Now multiplying out will give a term $-a^2x$. Again, we don't want one of those, and again we can get rid of it. This happens to give us the final expression; there is no remainder because, conveniently, $x - a$ is a factor of $x^3 - a^3$:

$$x^3 - a^3 = (x - a)(x^2 + ax + a^2).$$

This approach to polynomial division saves a lot of bother. If you want some practice with a specific numerical case, try dividing $x^4 - 9x^2 + 4x + 12$ by $x-2$ (then see if you can fully factorize the expression in the same way). What happens when we divide by a monomial that is not a factor? How do remainders come out?

Going back to our case, we can now write a general proof about the derivative of g.

Claim: If $g(x) = x^3$ then $g'(a) = 3a^2 \ \forall a \in \mathbb{R}$.

Proof: $\forall a \in \mathbb{R}$ we have $\dfrac{g(x) - g(a)}{x - a} = \dfrac{x^3 - a^3}{x - a} = x^2 + ax + a^2$.

So $\forall a \in \mathbb{R}, g'(a) = \lim\limits_{x \to a}(x^2 + ax + a^2) = 3a^2$.

Notice how I've handled the generality, both in the claim and in the proof. Can you think of other ways to do it? Notice also that we could add another line saying 'In particular, $g'(0) = 3 \cdot 0^2 = 0$'.

A general result about higher powers of n can be formulated as a theorem:

Theorem: Suppose that $n \in \mathbb{N}$ and $f_n : \mathbb{R} \to \mathbb{R}$ is given by $f_n(x) = x^n$.
Then $f_n'(x) = nx^{n-1}$.

This is often proved using induction on n and the product rule for differentiation. I will leave that for your course, but if you know about proof by induction you might like to have a go now.

I will conclude this section with a comment about meanings as they relate to graphs. What does it really mean to say that the derivative of $g(x) = x^3$ is $g'(x) = 3x^2$? Students often boggle when I ask this question, because they just 'know' the derivative, and either they haven't thought about the meaning for a long time or they haven't thought about it at all.

Thinking locally, it means that for any given point x, the gradient of the graph of g is captured by the value of the derivative. So, for instance, the gradient of g at -4 is $g'(-4) = 3.(-4)^2 = 48$ (a biggish positive number).

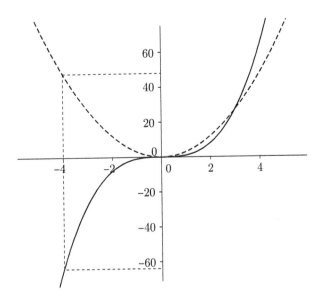

Thinking globally, it helps to imagine travelling along the graph of g from left to right. To start with, the graph slopes steeply upwards—this is reflected in the high positive value of g'. The slope of g gradually decreases until it is instantaneously zero—this is reflected in the graph of g' taking on the value zero. then the slope of g increases again, slowly at first but then faster—this is reflected in the fact that the value of g' increases again, slowly at first but then faster.

The fact that g and g' intersect at zero is a coincidence, of course. The graph of the function $h(x) = x^3 - 2$ would have the same derivative, so it might be useful to think about that one too. And what would be a similar description for, say, the function $f(x) = x^2$? I ask all this because students often know a bunch of derivatives but have forgotten what it all means, if they ever knew. If you didn't need the reminder, all the better.

8.5 Non-differentiability

The functions considered in Section 8.4 are differentiable and their derivatives are each expressible using a single formula. Thus it makes sense
to say things like 'the derivative of the function', treating differentiation
as a higher-level process that takes functions as inputs and returns other
functions as outputs. However, recall that the definition of differentiability does not apply to whole functions; it tells us whether or not a function
is differentiable *at a point*. This becomes more important when considering other functions, because many are differentiable at some points but
not at others.

The classic example, as noted in Section 8.1, is the function $f : \mathbb{R} \to \mathbb{R}$
given by $f(x) = |x|$. We can prove that this is not differentiable at 0 by
showing that the difference quotient tends to different limits depending
on the direction in which we approach 0 (the proof below involves the
notation '$x \to 0^+$', which would be read aloud as 'x tends to zero from
above').

Claim: $f : \mathbb{R} \to \mathbb{R}$ given by $f(x) = |x|$ is not differentiable at 0.

Proof: If $x > 0$ then $\dfrac{f(x) - f(0)}{x - 0} = \dfrac{|x| - |0|}{x - 0} = \dfrac{x}{x} = 1$.

So $\lim\limits_{x \to 0^+} \dfrac{f(x) - f(0)}{x - 0} = 1$.

If $x < 0$ then $\dfrac{f(x) - f(0)}{x - 0} = \dfrac{|x| - |0|}{x - 0} = \dfrac{-x}{x} = -1$.

So $\lim\limits_{x \to 0^-} \dfrac{f(x) - f(0)}{x - 0} = -1$.

$1 \neq -1$ so $\lim\limits_{x \to 0} \dfrac{f(x) - f(0)}{x - 0}$ does not exist.

So f is not differentiable at 0.

Did you experience a moment of confusion halfway through this proof,
on seeing $|x|$ replaced by $-x$? Many people do. This substitution occurs
in the case where $x < 0$, and it is valid because $|x|$ is defined formally
like this:

Definition: $|x| = \begin{cases} x & \text{if } x \geq 0 \\ -x & \text{if } x < 0 \end{cases}$.

If you haven't seen this before, check that it corresponds with your current understanding of $|x|$ (try $x = -2$, for instance), and read the proof again with this in mind.

While we are talking about $f(x) = |x|$, it is worth making a link back to Chapter 7. Recall that $f(x) = |x|$ is continuous at zero, which means (among other things) that it has a limit at zero. Doesn't this contradict what we just said? No, because we are talking about two different limits. When working with continuity, we consider the limit of the function values:

Definition: f is *continuous at a* if and only if $\lim\limits_{x \to a} f(x)$ exists and is equal to $f(a)$.

When working with differentiability, we consider the limit of the difference quotient:

Definition: f is *differentiable at a* if and only if $\lim\limits_{x \to a} \dfrac{f(x) - f(a)}{x - a}$ exists.

These are not the same limits. Make sure you keep this straight.

Also, note that it is correct in general to reason that if we get different 'gradients' as we approach a point from different sides, then the derivative at that point does not exist. However, this reasoning has the unfortunate side effect that it tends to make people believe that a function like this has derivative zero at the point a:

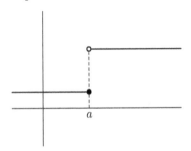

It's easy to see how this happens: the graph is 'flat' on both sides, so the derivative appears to be zero whether we approach *a* from the left or from the right. This is badly wrong, however, which we can see by labelling everything properly and relating this thinking more carefully to the definition. First, note that $f(a)$ is the lower of the two values, indicated in the standard way by a filled-in blob. If we label *a*, a point *x* to the right of *a*, and $f(x)$ and $f(a)$, we can see what really happens. As *x* approaches *a* from above, the gradient of the secant line tends to infinity—definitely not to zero.

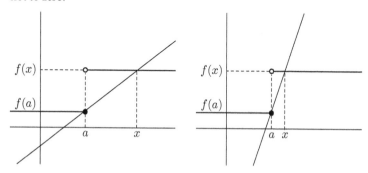

Here is a theorem relevant to this idea:

Theorem: If f is differentiable at a then f is continuous at a.

The contrapositive[4] of a true statement is always true, and the contrapositive of this theorem is the statement that if f is not continuous at a then f is not differentiable at a. The function in the diagrams above is not continuous at a so it can't be differentiable there.

In contrast, the converse of a true statement need not be true. The converse of this theorem would say that if f is continuous at a then f is differentiable at a. This is false: the function $f : \mathbb{R} \to \mathbb{R}$ given by $f(x) = |x|$ constitutes a counterexample. It is also worth recalling these functions from Chapter 7:

---

[4] The *contrapositive* of the conditional statement 'if A then B' is 'if not B then not A.' For a more detailed discussion of conditional statements, converses, inverses and contrapositives, see a transition-to-proof textbook or Section 4.6 of *How to Study for/as a Mathematics Degree/Major*.

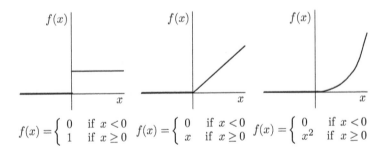

$$f(x) = \begin{cases} 0 & \text{if } x < 0 \\ 1 & \text{if } x \geq 0 \end{cases} \qquad f(x) = \begin{cases} 0 & \text{if } x < 0 \\ x & \text{if } x \geq 0 \end{cases} \qquad f(x) = \begin{cases} 0 & \text{if } x < 0 \\ x^2 & \text{if } x \geq 0 \end{cases}$$

The first function is not continuous at zero so it cannot be differentiable at zero. The second is continuous at zero but it is not differentiable there, and all students should be wary of such examples. It is common for a lecturer to present a function like this and ask students to state the derivative where it exists. Unwary students ignore the problem of differentiability and write this:

$$f'(x) = \begin{cases} 0 \text{ if } x < 0 \\ 1 \text{ if } x \geq 0 \end{cases}.$$

More attentive students realize that the function is not differentiable at zero and write this:

$$f'(x) = \begin{cases} 0 \text{ if } x < 0 \\ 1 \text{ if } x > 0 \end{cases}; \quad f \text{ is not differentiable at 0.}$$

Make sure you can see why the latter is correct.

What about the third function? This is both continuous at zero and differentiable at zero. Approaching zero from the left, both the function value and the difference quotient are equal to zero. Approaching from the right, both the function value and the difference quotient tend to zero. It would be a good idea to check this algebraically; doing so settles the questions raised at the end of Section 7.3.

8.6 Theorems involving differentiable functions

In Analysis you will encounter many theorems involving differentiability. Some will probably be presented together as a theorem like this:

Theorem (algebra for derivatives):
Suppose that $c \in \mathbb{R}$ and that $f : \mathbb{R} \to \mathbb{R}$ and $g : \mathbb{R} \to \mathbb{R}$ are differentiable at $a \in \mathbb{R}$.
Then

1. $f + g$ is differentiable at a with $(f + g)'(a) = f'(a) + g'(a)$ [*sum rule*];
2. cf is differentiable at a with $(cf)'(a) = cf'(a)$ [*constant multiple rule*].

Some students find it weird that we would write this down, because it seems obvious that $(f + g)'(a)$ and $f'(a) + g'(a)$ are the same. But this is really a theorem about *order of operations*. In $(f + g)'(a)$, the functions are added and then the result is differentiated. In $f'(a) + g'(a)$, the functions are differentiated and then the results are added. In mathematics more broadly, it is not obvious that orders of operations can be switched without changing the result. This theorem says that, provided all the derivatives are sensibly defined, switching works for derivatives.

I like to use this theorem to shake students out of any complacency. In a lecture, I write down the two parts above, then add

3. fg is differentiable at a with $(fg)'(a) = \ldots$.

Then I ask the class to say out loud what goes after the 'equals'. Practically everyone says '$f'(a)g'(a)$'. This is wrong, of course. As you have known for years, the *product rule* actually says

3. fg is differentiable at a with $(fg)'(a) = f(a)g'(a) + g(a)f'(a)$.

This wakes everyone up.

Proving the product rule involves a good trick,[5] but the proofs of all these results involve working directly with the definition, and they are not logically demanding so I will not devote space to them here. Once they are established, however, they can be used to prove that every polynomial function is differentiable everywhere. You might like to think about how such a proof would go.

---

[5] See Section 6.6 of *How to Study for/as a Mathematics Degree/Major*.

Here, we will look at some theorems that lead to the bigger results of Analysis, starting with Rolle's Theorem and the Mean Value Theorem (often abbreviated as MVT).

Rolle's Theorem:
Suppose that $f : [a, b] \to \mathbb{R}$ is continuous on $[a, b]$ and differentiable on (a, b), and that $f(a) = f(b)$. Then $\exists c \in (a, b)$ such that $f'(c) = 0$.

Mean Value Theorem:
Suppose that $f : [a, b] \to \mathbb{R}$ is continuous on $[a, b]$ and differentiable on (a, b). Then $\exists c \in (a, b)$ such that $f'(c) = \dfrac{f(b) - f(a)}{b - a}$.

Rolle's Theorem was discussed in Section 2.6. Don't turn back yet, though—first try to draw a diagram showing what each theorem says. For the Mean Value Theorem, it might not seem obvious how to represent the conclusion, but if you label everything properly you should be able to work out what the theorem says and to see intuitively why it must be true (I will explain below, but better if you try this yourself).

Once you've done that, one thing to notice is that Rolle's Theorem is a *special case* of the Mean Value Theorem—it is the case in which $f(a) = f(b)$, so $f(b) - f(a) = 0$. For that reason, we usually prove the Mean Value Theorem using a clever trick to reduce it to Rolle's Theorem. Here is the Mean Value Theorem again, along with a standard proof. Read it now, applying the self-explanation training from Section 3.5 (review that training first if necessary—remember that people understand more when they apply it properly).

Mean Value Theorem:
Suppose that $f : [a, b] \to \mathbb{R}$ is continuous on $[a, b]$ and differentiable on (a, b).
Then $\exists c \in (a, b)$ such that $f'(c) = \dfrac{f(b) - f(a)}{b - a}$.

> **Proof:** Assume that f is continuous on $[a, b]$ and differentiable on (a, b).
>
> Define $d : [a, b] \to \mathbb{R}$ by
> $$d(x) = f(x) - \left[f(a) + \left(\frac{f(b) - f(a)}{b - a} \right)(x - a) \right].$$

Now $f(a) + \left(\dfrac{f(b) - f(a)}{b - a} \right)(x - a)$ is a polynomial in x.

So d is continuous on $[a, b]$ and differentiable on (a, b) by the sum and constant multiple rules for continuous and differentiable functions.

Note that $d'(x) = f'(x) - \dfrac{f(b) - f(a)}{b - a}$.

Also $d(a) = f(a) - \left[f(a) + \left(\dfrac{f(b) - f(a)}{b - a} \right)(a - a) \right] = 0$

and $d(b) = f(b) - \left[f(a) + \left(\dfrac{f(b) - f(a)}{b - a} \right)(b - a) \right] = 0$.

So Rolle's Theorem applies to d on $[a, b]$.

So $\exists\, c \in (a, b)$ s.t. $d'(c) = 0$, that is s.t. $f'(c) - \dfrac{f(b) - f(a)}{b - a} = 0$.

So $\exists\, c \in (a, b)$ s.t. $f'(c) = \dfrac{f(b) - f(a)}{b - a}$ as required.

If you read the proof carefully, you will have noticed that the algebraic parts are actually quite simple. Because a, b, $f(a)$ and $f(b)$ are all just numbers, lots of things are constant, which means that differentiating the function d turns out to be straightforward. You might, however, have found the introduction of d a bit mystifying. Lots of students do, because d looks complicated and it seems to come out of nowhere. If you think about the proof in a more global way, though, you should see that d is used for the clever trick:[6] converting to the function d allows us to apply Rolle's Theorem; converting back gives the desired result about f. To apply Rolle's Theorem we need to establish that all its premises are satisfied, and the proof justifies this explicitly and thereby establishes the connection between the two theorems. To prove Rolle's Theorem, of course, would require different work—I will leave that for your course.

Thinking like this about logic and algebra is one way to understand a proof. But I think it's also satisfying to see why it works using diagrams. To do that for the MVT, notice that the expression $(f(b) - f(a))/(b - a)$, which appears in the theorem, is the gradient of the line joining the points

[6] Don't worry if you wouldn't have invented a trick like this yourself—your job as an Analysis student is to understand and adapt clever ideas that appear in standard proofs.

$(a, f(a))$ and $(b, f(b))$. So the theorem says that, provided the premises hold, there is a point c between a and b where the gradient of f is equal to the gradient of this line.

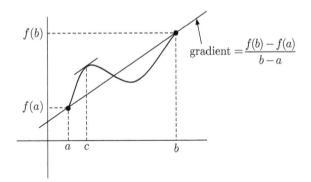

Also, the equation

$$y = f(a) + \left(\frac{f(b) - f(a)}{b - a} \right)(x - a)$$

is that of the straight line that passes through $(a, f(a))$ and $(b, f(b))$ (it is worth spending a few minutes working out why). So $d(x)$ gives the vertical difference between $f(x)$ and this line, and we can sketch what d would look like for a given function f. Doing so makes it clear that at points where the graph of f crosses the line, the value of d is zero; this occurs in particular at the endpoints a and b.

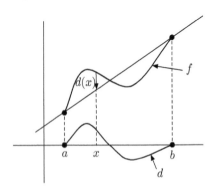

I suggest that you now read the proof again, referring to these diagrams to augment your self-explanations.

As you will have gathered, I like the insight that comes from linking logical arguments with diagrams. I like these diagrams especially because they allow me to see not only that the theorem must be true, but also how the proof works. Some theorems and proofs do not lend themselves to this type of reasoning—it is usually difficult to draw diagrams for proofs by contradiction, for instance, because they necessarily involve working from an incorrect assumption. And you might not like diagrams as much as I do. But I think it is often worth a shot.

To conclude this section we will examine some cute applications of the MVT. The theorem allows us, for instance, to establish things like this:

Theorem: Suppose that $f : \mathbb{R} \to \mathbb{R}$ is differentiable and that $\forall x \in \mathbb{R}$, $f'(x) = 0$. Then f is a constant function.

Stop and think about this for a moment. It does *not* say that if the function is constant, then the gradient is always zero. That would be easy to prove directly from the definition (think about how). This theorem is the converse[7] of that claim: it says that if the gradient is zero, then the function is constant. It is probably obvious to you that this must be the case. But for most people it is not obvious how to prove it, in part because constancy is a global property and it is not easy to argue from gradients to function values. The MVT provides a way to get a handle on this by relating function values to each other via derivatives at intermediate points.

It helps to rewrite the MVT like this, in order to focus attention on the difference between $f(a)$ and $f(b)$ (you can probably see how that will help):

Mean Value Theorem:
Suppose that $f : [a, b] \to \mathbb{R}$ is continuous on $[a, b]$ and differentiable on (a, b). Then $\exists c \in (a, b)$ such that $f(b) - f(a) = (b - a)f'(c)$.

Here is the theorem about constancy again, with a proof. Explain it to yourself, and think about how you would describe the overall strategy to another student.

[7] Section 2.9 discusses conditional statements, their converses and related logical issues.

Theorem: Suppose that $f : \mathbb{R} \to \mathbb{R}$ is differentiable and that $\forall x \in \mathbb{R}$, $f'(x) = 0$. Then f is a constant function.

Proof: Consider $a \in \mathbb{R}$ and suppose that $x \in \mathbb{R}$ with $x > a$.

Then f is differentiable on (a, x).

Also, because f is differentiable $\forall x \in \mathbb{R}$, f must be continuous on $[a, x]$ because differentiability at a point implies continuity at that point.

So, by the MVT, $\exists c \in (a, x)$ such that $f(x) - f(a) = (x - a)f'(c)$.

But $f'(c) = 0$ by the theorem premise.

So $\forall x > a, f(x) = f(a)$.

A similar argument proves that $\forall x < a, f(x) = f(a)$.

So $\forall x \in \mathbb{R}, f(x) = f(a)$, i.e. f is a constant function.

This proof invokes but does not spell out a 'similar argument' for the $x < a$ case. Invoking similar arguments is pretty common and is usually done when the author is confident that the reader would be able to fill in the missing steps. As a student, it's usually worth trying—doing so provides another way to make sure that you understand how a proof works.

In this case, a similar whole proof can be constructed to show that if the gradient of a function is always positive, then the function must be increasing. You might like work out how.

8.7 Taylor's Theorem

This final main section is about Taylor's Theorem, which is one of those things that students find difficult—it involves a lot of notation and some rather long equations so it looks intimidating, and people often try to avoid it. However, when thought about in the right way, it is not that complicated and it says something marvellous. In this section I want to make sure that you understand it, so that you can appreciate its value when it appears in your course.

To understand Taylor's Theorem it helps to understand the notion of a Taylor polynomial. Suppose we have a function $f : \mathbb{R} \to \mathbb{R}$ and a fixed point of interest a. Then the Taylor polynomial of degree n for f at a is

$$T_n[f,a](x) =$$

$$f(a)+f'(a)(x-a)+\frac{f''(a)}{2!}(x-a)^2+\frac{f^{(3)}(a)}{3!}(x-a)^3+\ldots+\frac{f^{(n)}(a)}{n!}(x-a)^n.$$

You see what I mean about the notation looking complicated. It isn't, though. In fact, each term in the polynomial has the same form, because $f^{(n)}(a)$ means the nth derivative[8] of f at a. Make sure you can see the pattern.

Look more thoughtfully and you will also see that the formula gives a polynomial in x, and that this polynomial has degree n. That's because lots of things are constant: a is a constant, which means that $f(a)$ is a constant, and so is $f'(a)$, and so is $f''(a)$, and so on. So the whole thing is just a bunch of constants multiplied by powers of x, and the highest power of x is n. This means that $T_n[f,a]$ is a function of x: for each value of $x \in \mathbb{R}$, we could calculate the values of all the terms and add them up, and $T_n[f,a](x)$ would vary as x varies.

So the Taylor polynomial has a fairly simple structure, but why is it interesting? It is interesting because it allows us to approximate a function f with a polynomial. I will demonstrate what that means using the function $f : \mathbb{R} \to \mathbb{R}$ given by $f(x) = \cos x$ and the fixed point $a = 2\pi/3$. We will start with Taylor polynomials where n is small.

The Taylor polynomial of degree one is

$$T_1[f,a](x) = f(a) + f'(a)(x-a).$$

Substituting in $f(x) = \cos x$ and $a = 2\pi/3$ gives:

$$T_1\left[\cos, \frac{2\pi}{3}\right](x) = \cos\left(\frac{2\pi}{3}\right) - \sin\left(\frac{2\pi}{3}\right)\left(x - \frac{2\pi}{3}\right)$$

$$= -\frac{1}{2} - \frac{\sqrt{3}}{2}\left(x - \frac{2\pi}{3}\right).$$

Students tend to multiply out the last expression, but I would advise against that. Usually when working with Taylor polynomials we want to

---

[8] People are often careless when they write this, omitting the brackets and just writing things like $f^3(a)$. But $f^3(a)$ is $f(a)$ cubed, and $f^{(3)}(a)$ is the third derivative of f at a. These are not the same thing at all—as ever, precision is important.

keep the structure visible, and a person reading this version will be able to see how it relates to the general formula.

Finding a Taylor polynomial is often straightforward because it just involves differentiation and substitution. But what does it mean? Graphs can help here. Plotting the graphs of f and $T_1[\cos, 2\pi/3]$ shows that the Taylor polynomial of degree one at the point $a = 2\pi/3$ is the tangent to f at $a = 2\pi/3$:

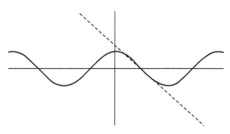

In fact, the Taylor polynomial of degree one is always the tangent at the point a. For the general case, one way to see this is to rearrange

$$T_1[f, a](x) = f(a) + f'(a)(x - a) \quad \text{to give} \quad f'(a) = \frac{T_1[f, a](x) - f(a)}{x - a},$$

which highlights the relationship between the value of $T_1[f, a]$ at x and the gradient of f at a:

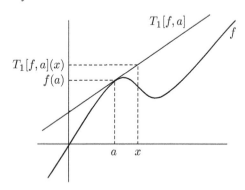

I also find it useful to observe (informally) that $T_1[f, a]$ 'matches' the graph of f in the sense that it has the same value at a and the same derivative at a.

What do you think happens for the second order Taylor polynomial? It 'matches' the graph of f in the sense that it has the same value at a and the same derivative at a and the same second derivative at a. The general second order Taylor polynomial is

$$T_2[f,a](x) = f(a) + f'(a)(x-a) + \frac{f''(a)}{2!}(x-a)^2,$$

and substituting in $f(x) = \cos x$ and $a = 2\pi/3$ (again keeping the structure visible rather than multiplying out) gives:

$$T_2\left[\cos,\frac{2\pi}{3}\right](x) = \cos\left(\frac{2\pi}{3}\right) - \sin\left(\frac{2\pi}{3}\right)\left(x - \frac{2\pi}{3}\right)$$
$$- \cos\left(\frac{2\pi}{3}\right)\left(x - \frac{2\pi}{3}\right)^2$$
$$= -\frac{1}{2} - \frac{\sqrt{3}}{2}\left(x - \frac{2\pi}{3}\right) + \frac{1}{2}\left(x - \frac{2\pi}{3}\right)^2.$$

Plotting the graphs this time gives:

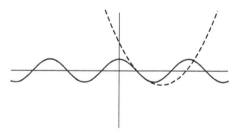

You can probably guess what happens if we keep going. Here is the graph with the Taylor polynomial of degree 3:

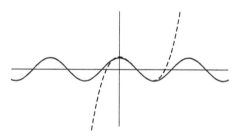

And here is the graph with the Taylor polynomial of degree 30.

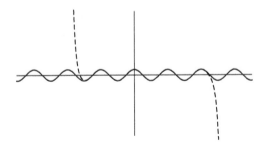

It turns out that we can get as good an approximation as we like, as far away from a as we like, by taking more and more terms. If we could take infinitely many terms, the graphs would match perfectly. I think you will agree that this is pretty cool. And if you are aware of it then you are in a good position to understand Taylor's Theorem.

Taylor's Theorem:
Let I be an open interval containing a and x. Suppose that f is $(n+1)$ times differentiable on I and that $f^{(n+1)}$ is continuous on I. Then there exists c between a and x such that

$$f(x) = T_n[f, a](x) + \frac{f^{(n+1)}(c)}{n!}(x - c)^n(x - a).$$

Again this looks intimidating, but if we look past all the notation it has this structure:

Taylor's Theorem: Let a bunch of conditions[9] hold. Then

$$f(x) = \text{Taylor polynomial} + \text{another bit}.$$

The other bit is often referred to as a *remainder term*, which makes sense because the theorem says that the function value is equal to the Taylor polynomial of degree n plus whatever is left over. Looking more carefully

---

[9] These conditions are sensible. For instance, we need the function to be differentiable $n + 1$ times if we want the derivatives in the formula to exist.

at the remainder term reveals that the remainder will be small if n is large and $x - a$ is small (which forces $x - c$ to be small—why?). In other words, the approximations are better for x close to a and for big values of n. So the theorem should seem reasonable given the material you have read in this section.

In fact, Taylor's Theorem can be formulated in a number of ways with slightly different expressions for the remainder term. But all the expressions share these properties, meaning that the theorem can be thought of as telling us about approximating functions with polynomials by making the remainder term small. Keep this in mind and you should find that work on Taylor's Theorem makes intuitive sense.

8.8 Looking ahead

In a typical Analysis course, work on differentiability will cover the material from this chapter, with more examples and with proofs of all the theorems. Studying differentiability is particularly good for observing the buildup of theory (as discussed in Section 3.2): the Extreme Value Theorem (see Section 7.11) is used to prove Rolle's Theorem, Rolle's Theorem is used to prove the Mean Value Theorem, and the Mean Value Theorem is used to prove Taylor's Theorem. Indeed, this book's focus means that I have taken a top-down approach, stating theorems then explaining how to understand them, but your lecturer might take a bottom-up approach instead so that the relevant reasoning is done first and the theorems appear as its natural consequences.

Either way, you will also find that the sum and product rules for continuous and differentiable functions pop up all over the place. And you might apply some or all of these ideas in study of the chain rule (which you already know), L'Hôpital's rule (which you might have come across in calculus), and ways of identifying local maxima and minima when the second derivative test yields no information. You might also examine Taylor series for specific functions about specific points, and perhaps functions for which Taylor polynomials do not give good approximations.

In courses on multivariable calculus, you will generalize the ideas of differentiability and derivatives to functions of two or more variables—try thinking now about what differentiability ought to mean for a function that defines a surface instead of a curve. And in vector calculus you will learn about applying these ideas in different coordinate systems, and in the solution of partial differential equations.

Back in Analysis, you will also learn about links between differentiability and integrability. For differentiable functions, this link sounds straightforward—differentiation and integration are inverse operations. But what does that really mean? And what happens for non-differentiable functions? These questions require proper consideration of the meaning of integrability, and they are addressed in the next chapter.

CHAPTER 9

Integrability

This chapter discusses the concept of integrability, as distinct from the process of integration. It examines relationships between antiderivatives and areas under graphs, then builds up the definition of integrability via approximations to areas. It gives an example of a function that is not integrable, and shows how Riemann's condition—a test for integrability—can be used in proofs. It concludes by explaining the Fundamental Theorem of Calculus.

9.1 What is integrability?

This chapter is about *integrability*, not about integration. That's because Analysis, as discussed in Section 3.3, is about the theory underlying earlier mathematics. For integrability it might not be obvious what this means. You might have been taught that integration is antidifferentiation—the 'opposite' of differentiation. That is not wrong (though it is better to say that differentiation and integration are *inverse operations*). And you might have been taught that integration is about finding the area under a graph. That is not wrong either. But it leaves open plenty of questions.

For a start, why should differentiation and integration be inverse operations? Why is finding a gradient or slope[1] the 'opposite' of finding an area under a graph? If a teacher said so and you just believed it, fair enough. But if you have never thought seriously about this relationship, you should do so now. Many people, when they do, agree that it is astonishing. Why on earth would gradients and areas be related?

[1] As noted in previous chapters, British people use the word 'gradient' and Americans use the word 'slope' to mean the same thing.

They seem to involve completely different concepts. But mathematics is not arbitrary—the relationship doesn't exist because some authority has randomly decreed that it should—so there must be a reason. And the fact that the reason is not obvious means that there must be some deep mathematics to learn.

For another thing, early study of differentiation and integration usually involves simple functions defined by single formulas. But Chapters 7 and 8 introduced functions with more complicated specifications, which raises the possibility that differentiation and integration might not always be inverse operations in an unproblematic way. Consider, for instance, the function shown below (you'll shortly see why I'm using t as the variable[2] rather than x).

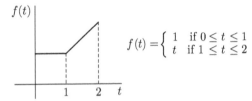

$$f(t) = \begin{cases} 1 & \text{if } 0 \leq t \leq 1 \\ t & \text{if } 1 \leq t \leq 2 \end{cases}$$

This function is not differentiable at $t = 1$. So integration can't be a straightforward 'opposite' of differentiation in this case. Nevertheless it seems sensible to talk about the area under the graph between 0 and 2, or between 0 and a general point $x \in [0, 2]$. That area can be calculated using the diagram and formula below (why did I use t instead of x?). Imagine x sliding around on the axis and use simple geometry to check that the formula is valid.

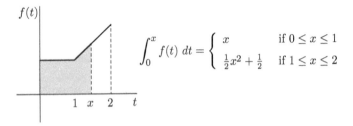

$$\int_0^x f(t)\, dt = \begin{cases} x & \text{if } 0 \leq x \leq 1 \\ \frac{1}{2}x^2 + \frac{1}{2} & \text{if } 1 \leq x \leq 2 \end{cases}$$

[2] This is fine because the function given by $f(t) = 3t$ is the same as the function given by $f(x) = 3x$ or $f(j) = 3j$ or whatever. We often use standard letters for standard things because it helps everyone to grasp new ideas more quickly, but we don't have to stick to a particular notation if there is a good reason to use a different one.

How does this result relate to straightforward formulaic integration for the function f? Try it and see. Many students assume that piecewise-defined functions can be integrated simply by integrating both pieces, but this is not the case. If you have read Section 8.6 you will be aware that such a function might not be differentiable at the 'join'; the piecewise nature of this function does not disrupt integrability, but we do need to be cautious about constants of integration.

There are, however, functions for which integrability is a problem, like this one:

$$f : \mathbb{R} \to \mathbb{R} \text{ given by } f(x) = \begin{cases} 1 \text{ if } x \in \mathbb{Q} \\ 0 \text{ if } x \notin \mathbb{Q} \end{cases}.$$

Many people agree that it does not make sense to think about the 'area under the graph' for this function. So we should expect a definition of integrability that classifies this function as non-integrable. We will check that in Section 9.5.

9.2 Areas and antiderivatives

Before defining integrability, I want to unpack the relationship between antiderivatives and areas under graphs. We tend to talk about a function having 'an' antiderivative, saying things like 'the antiderivative of x^2 is $x^3/3 + c$'. But that is not a single function, it is an infinite family of functions, one for each value of c. Perhaps this is good because under a particular graph there is not just one area—the area under the graph from a to x will not generally be the same as the area under the graph from a different number b to x:

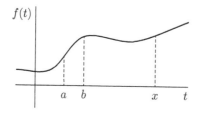

So how exactly do antiderivatives and areas fit together? You might have thought about this before, but many new Analysis students have spent more time on calculations than on conceptual understanding and they aren't able to give a good answer. Pause now and see whether you can, then read on.

To sort out the relationship, we will start by finding areas. Consider, for instance, the simple function $f : \mathbb{R} \to \mathbb{R}$ given by $f(t) = 3t$, for which the integral from 0 to a general point x is just the area of a triangle:

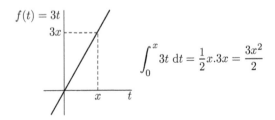

$$\int_0^x 3t \ \mathrm{d}t = \frac{1}{2}x.3x = \frac{3x^2}{2}$$

The integral thus matches what we find using antiderivatives. But finding the area gives just one antiderivative—the one where the constant happens to equal 0. What happens if, instead of starting at zero, we integrate f between a different fixed number a and the variable number x? Keeping everything positive and $x > a$ for simplicity gives this:

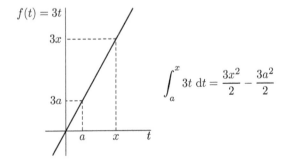

$$\int_a^x 3t \ \mathrm{d}t = \frac{3x^2}{2} - \frac{3a^2}{2}$$

Starting at a instead of 0 chops off a bit of area. But this bit of area is always the same, so the effect is to adjust the integral by a constant (what happens if a is negative?). That explains why an antiderviative should be a family of functions that differ only by a constant.

An alternative is to think dynamically about what happens as x varies. Moving from a fixed x_1 to a fixed x_2 adds a fixed amount of area—it doesn't matter where we start, we always collect the same amount of area between these points. This means that at any particular point, the *rate of change* of the growing area is fixed. The rate is not the same everywhere—in the diagram below, the amount of area added between x_3 and x_4 is bigger than that added between x_1 and x_2. But the rate of change at x_1 is a meaningful quantity, as is the (bigger) rate of change at x_3.

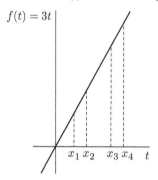

This means that we can think not only about an area under a graph but also about the *rate of change* of that area with respect to x. Rates of change, of course, are modelled using derivatives. So it is starting to seem reasonable that areas and derivatives should be closely related. We will establish exactly how in Section 9.8, after building up the appropriate definitions.

9.3 Approximating areas

The areas under the graphs in Section 9.2 have simple shapes, and we want to be able to work with more complicated functions, including those with curved graphs. If your first thought is that we should use integration, be careful. That leads to a circular argument because it amounts to saying that we find integrals by calculating areas and we find areas by calculating integrals. It is philosophically important, therefore, to sort out the question of area measurement, and to recognize that although we have an intuitive sense of what it means to talk about the area under a curved graph, obtaining a numerical measure of that area is not trivial—it cannot be done by straightforward multiplication.

Mathematicians get around this problem by working with approximations. For $\int_a^b f(x)dx$, they consider estimates like those in the diagrams below, where the total areas of the rectangles provide an underestimate (on the left) and an overestimate.

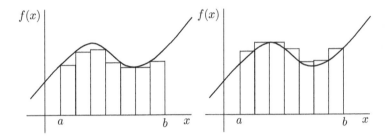

Narrower rectangles (in general) give better approximations:

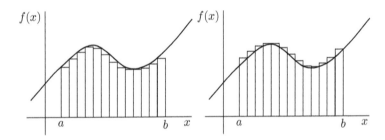

No approximation gives the exact area, but the idea is that if both the underestimates and the overestimates can be made as close as we like to a particular number A, then A is the area under the graph. Similar reasoning is used everywhere in Analysis—we think in terms of getting as close as we like to a limiting value. Here this gets us out of the philosophically problematic circular reasoning: mathematicians start with areas that are meaningful—those of the rectangles—and use them to *define what area means* for a shape with a curved edge.[3]

---

[3] You might know more sophisticated ways of approximating integrals using, for instance, the trapezium rule or Simpson's rule. But an approach using rectangles is algebraically simpler and works just fine, so this is likely to be the one you meet first in Analysis.

Most students find this business with rectangles and approximations intuitively reasonable. So all we need to do is formalize it. Unfortunately, doing so involves introducing a lot of notation, which means that many students end up thinking integrability is hard. But it isn't hard. Indeed, early work with integrability is logically simpler than early work with continuity (for instance). So I hope to convince you that the notation just captures the intuitive ideas outlined above.

The first step in formalizing is to think about integrability on a restricted domain. If you have read Chapters 7 and 8, this will not be a surprise—both continuity and differentiability are first defined at a point. It doesn't make sense to talk about integrability at a point, but the same general idea applies because there exist functions that are integrable on some parts of the number line but not on others. So mathematicians usually describe a function as integrable or not on an interval $[a, b]$. With that established, the strategy is to construct:

- an expression for how the interval $[a, b]$ is split up;
- an expression for the area of a single rectangle;
- an expression for a single overestimate, formally called an *upper sum* (what do you think an underestimate is called?);
- an expression specifying the area A by relating it to the upper sums;
- a definition of integrability on the interval.

Each of these steps introduces notation, and students sometimes get confused because they have failed to understand a particular step or have muddled two steps up. After the formalization, I will repeat the list so that you can review the whole construction.

9.4 Integrability definition

Splitting up the interval $[a, b]$ usually involves calling the leftmost point x_0, then labelling the others in the obvious way and calling the last one x_n. You might see the definition below; in the accompanying diagram, $n = 5$ and the total area of the rectangles gives the overestimate associated with the partition $\{x_0, \ldots, x_5\}$.

Definition: A *partition* of the set $[a, b]$ is a finite set[4] of points $\{x_0, \ldots, x_n\}$ such that $a = x_0 < x_1 < \ldots < x_{n-1} < x_n = b$.

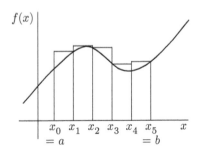

Does the definition of *partition* specify that all the subintervals have equal widths? No. You will often see mathematical arguments that do make them equal, but this is only for computational convenience.

Finding the areas of the rectangles is straightforward. The first one has width $x_1 - x_0$, and this will always be the case. In the diagram above, its height happens to be $f(x_1)$, but that need not always be the case—look at some of the others to see why. For a general subinterval $[x_{j-1}, x_j]$, the height is the largest value of $f(x)$ on the subinterval.[5] We often give this height a name, saying something like

Let $M_j = \sup\{f(x) | x_{j-1} \le x \le x_j\}$.

Here 'sup' is an abbreviation for *supremum*, so this can be read aloud as 'Let M_j be equal to the supremum of the $f(x)$ values for x between x_{j-1} and x_j'. A definition of *supremum* can be found in Section 10.5; here you can continue to think of M_j informally as the largest value of f on the subinterval $[x_{j-1}, x_j]$, although this is a bit imprecise relative to the definition (read Section 10.5 to find out why). In any case, the area of the first rectangle is $M_1(x_1 - x_0)$. Where would you mark M_1, M_2, M_3, M_4 and M_5

---

[4] You'd be amazed how many people lose marks on exams because they write 'a set of finite points' instead of 'a finite set of points'. All points are finite so the former doesn't really say anything. As ever, pay attention to the detail.

[5] This assumes that the function does have a largest value on the subinterval, which means the function must be bounded. Sometimes the requirement of boundedness is incorporated into the definitions associated with integrability.

on the vertical axis of the diagram? What are the areas of the other rectangles? And why is it important that the definition of M_j involves '$\leq$' rather than '$<$' symbols?

Having found areas for the rectangles, we want to add them up to get the associated overestimate for $\int_a^b f(x)\mathrm{d}x$. The overestimate is denoted '$U(f;P)$', which is read aloud as 'the upper sum for f with respect to the partition P' and which can be found using this formula:[6]

$$U(f;P) = \sum_{j=1}^{n} M_j(x_j - x_{j-1}), \text{ where } M_j = \sup\{f(x)|x_{j-1} \leq x \leq x_j\}.$$

Remember that when you see sigma notation it is often a good idea to write out the expression in full. For our diagram this gives the expected sum

$$U(f;P) = \sum_{j=1}^{5} M_j(x_j - x_{j-1})$$
$$= M_1(x_1 - x_0) + M_2(x_2 - x_1) + M_3(x_3 - x_2) + M_4(x_4 - x_3) + M_5(x_5 - x_4).$$

An equivalent general expression for the underestimate, called the *lower sum*, is

$$L(f;P) = \sum_{j=1}^{n} m_j(x_j - x_{j-1}) \text{ where } m_j = \inf\{f(x)|x_{j-1} \leq x \leq x_j\}.$$

Here 'inf' is an abbreviation for *infimum* (again, see Section 10.5).

Notation for this kind of thing varies a bit; your lecturer or textbook might use a different notation for $U(f;P)$, for instance. Also, it is possible to specify the rectangles in different ways. But the principles are always similar, so you should be able to work out how these ideas (and others in this chapter) are captured by any variants in your course.

In any case, these definitions specify a single upper sum or a single lower sum. And a single upper sum is just a number; the formula captures a lot of calculation but it yields a single total area. It is important to keep

---

[6] '$\sum$' is the upper-case Greek letter 'sigma'—see Section 6.2 if sigma notation is unfamiliar or if you need a review.

track of this because there are many different possible partitions, each of which will have its own upper sum. Maybe one upper sum is 17, another is 18, another is 18.5, and so on. These upper sums are all approximations to the area under the graph, but how does that area relate to them? First, the integral A will be less than or equal to every upper sum. Second, it will be the greatest number with that property. In other words it will be the *greatest lower bound* for the set of all possible upper sums, also known as the *infimum* of the upper sums (again, see Section 10.5):

$$A = \inf\{U(f; P) | P \text{ is a partition of } [a, b]\}.$$

How should A be related to all the lower sums and what would we write?

Finally, all of this reasoning assumes that the area A is meaningful. But it is only meaningful if we get the same value for A regardless of whether we use the lower or the upper sums. Hence the definition of integrability:

Definition: f is *integrable* on the interval $[a, b]$ if and only if

$\inf\{U(f; P) | P \text{ is a partition of } [a, b]\} = \sup\{L(f; P) | P \text{ is a partition of } [a, b]\}$.

To conclude this section, here is the promised recap of the list of constructions. Can you draw a diagram and reconstruct all the expressions without needing to look?

- an expression for how the interval $[a, b]$ is split up;
- an expression for the area of a single rectangle;
- an expression for a single overestimate, formally called an *upper sum*;
- an expression specifying the area A by relating it to the upper sums;
- a definition of integrability on the interval.

9.5 A non-integrable function

In Section 9.1 I observed that it does not seem meaningful to talk about the area under the graph for the function below (the graph cannot be sketched accurately but, as discussed in Section 7.3, the dotted lines provide some intuition).

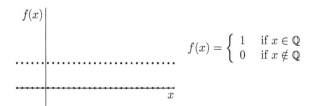

$$f(x) = \begin{cases} 1 & \text{if } x \in \mathbb{Q} \\ 0 & \text{if } x \notin \mathbb{Q} \end{cases}$$

This section explains how this claim relates to the formal definition. Before reading on, try to anticipate the argument by thinking about how integrability is defined. At which step will we run into trouble?

To consider integrability we need an interval $[a, b]$ (and we'll assume $a \neq b$ or there is nothing to talk about). Suppose we split up the interval using the partition $a = x_0 < x_1 < \ldots < x_{n-1} < x_n = b$. What are the associated lower and upper sums? The upper sum is defined as

$$U(f; P) = \sum_{j=1}^{n} M_j(x_j - x_{j-1}) \text{ where } M_j = \sup\{f(x)|x_{j-1} \leq x \leq x_j\},$$

and every subinterval contains rational numbers so every M_j must equal 1. This means that a bunch of things cancel when we expand $U(f; P)$:

$$U(f; P) = 1(x_1 - x_0) + 1(x_2 - x_1) + \ldots + 1(x_{n-1} - x_{n-2}) + 1(x_n - x_{n-1})$$
$$= x_n - x_0$$
$$= b - a.$$

Because there was nothing special about the partition, every other partition also yields the upper sum $b - a$. So the greatest lower bound of the upper sums is $b - a$. In symbols,

$$\inf\{U(f; P)|P \text{ is a partition of } [a, b]\} = b - a.$$

The lower sum of f with respect to P is defined as

$$L(f; P) = \sum_{j=1}^{n} m_j(x_j - x_{j-1}) \text{ where } m_j = \inf\{f(x) | x_{j-1} \leq x \leq x_j\}.$$

This time, every m_j must equal 0. Why? And why does this mean that

$$\sup\{L(f; P) | P \text{ is a partition of } [a, b]\} = 0?$$

Finally, what does all this mean in terms of the integrability definition? The infimum of the upper sums is $b - a$ and the supremum of the lower sums is zero, so

$$\inf\{U(f; P) | P \text{ is a partition of } [a, b]\} \neq \sup\{L(f; P) | P \text{ is a partition of } [a, b]\}$$

meaning that f is not integrable on $[a, b]$.

Do you think that a similar argument could be used to show that every non-continuous function must non-integrable? What if a function is discontinuous at only one point? We'll look at a case like that in the next section.

9.6 Riemann's condition

If your course defines integrability as above, the definition might say not just 'f is integrable' but 'f is Riemann[7] integrable'. This reflects the fact that there are other approaches to defining integrability and integrals. These tend to come up in more advanced courses so I will not discuss them here, but I do want to draw your attention to something else that bears Riemann's name.

Theorem (Riemann's condition):
f is (Riemann) integrable on $[a, b]$ if and only if for every $\varepsilon > 0$ there exists a partition P of $[a, b]$ such that $U(f; P) - L(f; P) < \varepsilon$.

The expression '$U(f; P) - L(f; P) < \varepsilon$' is the difference between an upper sum and the corresponding lower sum, which in the following diagram is represented as the total area of the grey boxes.

[7] *Riemann* is a German name so it is pronounced 'Reeman'.

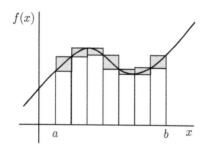

Informally, Riemann's condition says that a function is integrable if and only if, by considering different partitions, we can make this difference as small as we want. I won't prove here that Riemann's condition is valid—though you should think about how it relates to the definition— but I will illustrate how it can be applied and highlight a detail that people sometimes miss when thinking about upper and lower sums. Consider the function $f : [0,2] \to \mathbb{R}$ specified below. For the partition $\left\{0, \frac{1}{3}, \frac{2}{3}, 1, \frac{4}{3}, \frac{5}{3}, 2\right\}$, what are the upper and lower sums, and what is the difference between them?

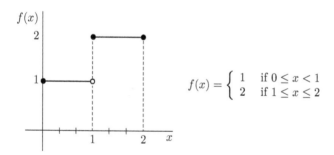

$$f(x) = \begin{cases} 1 & \text{if } 0 \leq x < 1 \\ 2 & \text{if } 1 \leq x \leq 2 \end{cases}$$

Did you say that the difference is zero? That's wrong. It's wrong in the obvious way so there is nothing weird about your thinking if that's what you said, but you have been misled by the overall appearance of the graph. Think again before you go on.

Below is a visual representation of the upper sum. If you said zero, can you see why that wasn't right? The key is the subinterval $\left[\frac{2}{3}, 1\right]$. The point 1 is included in this subinterval, and $f(1) = 2$, so $\sup\left\{f(x)\middle|\frac{2}{3} \le x \le 1\right\} = 2$ and $U(f; P) - L(f; P) = \frac{1}{3}$.

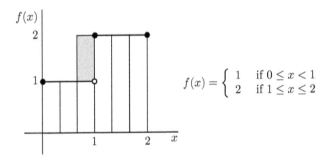

$$f(x) = \begin{cases} 1 & \text{if } 0 \le x < 1 \\ 2 & \text{if } 1 \le x \le 2 \end{cases}$$

Can we make $U(f; P) - L(f; P)$ as small as we like by using different partitions? The answer is yes, and you might like to think about how to write a full, convincing proof of this (full doesn't have to mean long). Thus Riemann's condition is satisfied and this function is integrable on the interval $[0, 2]$. So a function can be integrable without being continuous.

At a higher level, there is something to notice here about the structure of mathematical theory. I stated Riemann's condition as a theorem, and you will probably see it proved. However, the theorem has an if-and-only-if structure, meaning that the condition is logically equivalent to the definition of integrability. Technically, therefore, we could use Riemann's condition as the definition and prove the original definition as a theorem. When this kind of thing occurs, mathematicians make decisions about what to treat as fundamental and what to treat as a derivable result. Usually everyone agrees, but you might see variations for some concepts, perhaps in different textbooks. This does not mean that one source is wrong or outdated—it just reflects this type of logical equivalence relationship.

9.7 Theorems involving integrable functions

Much early work on integrability is of the type discussed in Section 3.2. It involves, for instance, proving that if f and g are both integrable on $[a, b]$, then so is $f + g$. Proofs of such things tend to look quite long, but they often just involve banging the appropriate information into the definitions of lower and upper sums and adding things up. Here is such a claim and proof for reading practice.

Claim: If f is integrable on $[a, b]$ then $3f$ is integrable on $[a, b]$.

Proof: Suppose that f is integrable on $[a, b]$ and let $\varepsilon > 0$ be arbitrary.

By Riemann's condition there exists a partition P of $[a, b]$ such that $U(f; P) - L(f; P) < \frac{\varepsilon}{3}$.

Now, by definition,

$U(3f; P) = \sum_{j=1}^{n} M_j(x_j - x_{j-1})$

$\qquad\qquad$ where $M_j = \sup\{3f(x) : x_{j-1} \leq x \leq x_j\}$ and

$L(3f; P) = \sum_{j=1}^{n} m_j(x_j - x_{j-1})$

$\qquad\qquad$ where $m_j = \inf\{3f(x) : x_{j-1} \leq x \leq x_j\}$.

Also, by general properties of suprema and infima, $\forall j \in \{1, \ldots n\}$ we have

$\sup\{3f(x) : x_{j-1} \leq x \leq x_j\} = 3\sup\{f(x) : x_{j-1} \leq x \leq x_j\}$ and

$\inf\{3f(x) : x_{j-1} \leq x \leq x_j\} = 3\inf\{f(x) : x_{j-1} \leq x \leq x_j\}$.

So $U(3f; P) = 3U(f; P)$ and $L(3f; P) = 3L(f; P)$.

So $U(3f; P) - L(3f; P) = 3(U(f; P) - L(f; P)) < 3\frac{\varepsilon}{3} = \varepsilon$.

So $3f$ satisfies Riemann's condition on $[a, b]$.

Hence $3f$ is Riemann integrable on $[a, b]$.

As usual, you should think about how this claim and proof might generalize. What if 3 were replaced by 6, or by –3, or by a general constant c? Why is the result about general properties of suprema and infima[8] valid? And how might we prove that if f and g are both integrable on $[a, b]$, then so is $f + g$ (for inspiration, see the proof of the sum rule for convergent sequences in Section 5.10)?

Next, a theorem involving more concepts.

Theorem: Suppose that f is bounded and increasing on $[a, b]$.

Then f is Riemann integrable on $[a, b]$.

I like this theorem because a standard proof based on Riemann's condition can be captured very elegantly in a diagram, as below. The argument is that the difference between the upper and lower sums—the total area of the grey boxes—is the same as the area of the rectangle to the right. The area of the rectangle is $f(b) - f(a)$ times its width, so it can be made smaller than any particular ε by making the width small enough. So Riemann's condition is satisfied and the theorem must be true.

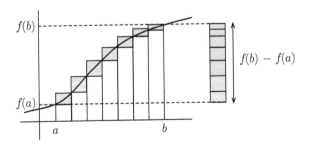

As you read the proof, think about how each part relates to the diagram.

---

[8] *Supremum* is the singular, *suprema* is the plural—this is like *maximum* and *maxima*.

Theorem: Suppose that f is bounded and increasing on $[a, b]$.

Then f is Riemann integrable on $[a, b]$.

Proof: Let $\varepsilon > 0$ be arbitrary.

Note that $f(b) - f(a) \geq 0$ because f is increasing on $[a, b]$.

Choose $N \in \mathbb{N}$ s.t. $\dfrac{b - a}{N}(f(b) - f(a)) < \varepsilon$.

Let P_N be the partition $\{x_0, x_1, \ldots, x_N\}$ with

$$x_j - x_{j-1} = \frac{b - a}{N} \quad \forall j \in \{1, \ldots, N\}.$$

Because f is increasing, $\forall j \in \{1, \ldots, N\}$ we have

$\sup\{f(x) : x_{j-1} \leq x \leq x_j\} = f(x_j)$ and
$\inf\{f(x) : x_{j-1} \leq x \leq x_j\} = f(x_{j-1})$.

So $U(f; P_N) = \displaystyle\sum_{j=1}^{N} f(x_j)(x_j - x_{j-1}) = \frac{b - a}{N} \sum_{j=1}^{N} f(x_j)$

and $L(f; P_N) = \displaystyle\sum_{j=1}^{N} f(x_{j-1})(x_j - x_{j-1}) = \frac{b - a}{N} \sum_{j=1}^{N} f(x_{j-1})$.

So $U(f; P_N) - L(f; P_N) = \dfrac{b - a}{N} \left(\displaystyle\sum_{j=1}^{N} f(x_j) - \sum_{j=1}^{N} f(x_{j-1}) \right)$

$$= \frac{b - a}{N} \left(f(x_N) - f(x_0) \right)$$

$$= \frac{b - a}{N} \left(f(b) - f(a) \right)$$

$$< \varepsilon.$$

So Riemann's condition is satisfied.

Hence f is integrable on $[a, b]$.

Would the conclusion of this theorem still hold if f were decreasing on $[a, b]$? If so, how would the proof need to change? And does the function in this theorem have to be continuous? Even when you've seen lots of non-continuous functions your brain will keep defaulting to continuous ones because they're more familiar. So do remind yourself every now and then to think beyond the obvious cases.

9.8 The Fundamental Theorem of Calculus

This chapter began with an informal discussion of the relationship between integration and differentiation. This relationship is captured formally by the Fundamental Theorem of Calculus (often abbreviated as 'FTC'). If you have studied a lot of calculus, you might have seen a proof of the FTC already. The proofs in calculus courses, however, tend to make numerous assumptions. Unsurprisingly, an Analysis course will establish everything properly on the basis of definitions and earlier theorems.

Analysis courses tend to go pretty fast, however, and many students rush through a proof of the FTC before they've really understood the theorem. Here I want to make sure that you understand both what the FTC says and what it would take to prove it.

Theorem (Fundamental Theorem of Calculus):

Suppose that f is integrable on $[a, b]$ and let $F(x) = \displaystyle\int_a^x f(t)dt$. Then

1. F is continuous on $[a, b]$;
2. If f is continuous on $[a, b]$ then F is differentiable on $[a, b]$ and $F'(x) = f(x)$.

The notation in the theorem premise can be understood by inspecting a diagram illustrating the relationship between f and F:

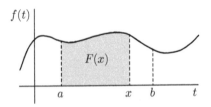

Using the same notation, it's possible to get a decent informal grasp on why integration and differentiation are inverse operations by thinking about approximations for a general function f and the associated F. From Section 8.4, the derivative of F at x is defined as follows.

Definition: $F'(x) = \lim\limits_{h \to 0} \dfrac{F(x + h) - F(x)}{h}$, provided this limit exists.

Then consider this diagram and argument. The approximation gets better as $h \to 0$, making it plausible that, in the limit, $F'(x) = f(x)$.

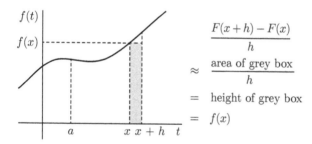

$$\dfrac{F(x + h) - F(x)}{h}$$
$$\approx \dfrac{\text{area of grey box}}{h}$$
$$= \text{height of grey box}$$
$$= f(x)$$

So the FTC is consistent with the claim that integration and differentiation are inverse operations. But the theorem clearly says something more precise and complex than this—its conclusions about the relationship between the function f and its integral F depend on the nature of f, which probably isn't something you've considered before.

To understand exactly what the FTC says I find it useful to consider a specific example such as the function $f : [0, 2] \to \mathbb{R}$ given by

$$f(t) = \begin{cases} 1 & \text{if } 0 \le t \le 1 \\ 2 & \text{if } 1 < t \le 2 \end{cases}.$$

Section 9.6 established that although this function is not continuous at 1, it is nevertheless integrable on $[0, 2]$. So the FTC applies. Considering the graph of the corresponding integral F can clarify the meanings of both parts of the FTC.

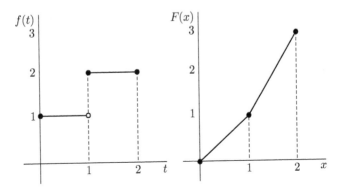

First, make sure you are convinced that F is shown correctly. I find it helpful to observe that on the interval $[0, 1]$, the area under the graph of f increases at a constant rate, and the integral F must reach 1 at $x = 1$. On the interval $[1, 2]$, the area under the graph of f also increases at a constant rate but now that rate is doubled, and the integral F must reach 3 at $x = 2$.

Then think about part 1 of the FTC, which simply says that if f is integrable then F is continuous. The graph of f has a 'jump' at $x = 1$, but this does not result in a corresponding jump in F because the area does not jump to a new value instantaneously as x passes through the value 1. So F is continuous even though f is not. This should help you to see why the part 1 of the FTC is reasonable.

Part 2 of the FTC says that if f is continuous then F is differentiable. In our example f is *not* continuous at 1 and F is *not* differentiable: the graph of F has a 'corner' at $x = 1$. This should clarify why the FTC needs the extra continuity condition in part 2: if f is not continuous then there might be sharp changes in the gradient/slope of F. Overall, then, this is an example in which differentiation and integration are not straightforward inverse operations. Because most functions in their earlier study of differentiation and integration are continuous everywhere, this thinking is new to most Analysis students.

Proving the FTC provides an opportunity to draw together the concepts of limits, continuity, differentiability and integrability. To prove part 1, that F is continuous, we need to prove that F satisfies the definition of continuity (see Sections 7.4 and 7.5). This means proving that for every $c \in [a, b]$ it is true that

$$\forall \varepsilon > 0 \, \exists \delta > 0 \text{ such that if } |x - c| < \delta \text{ then } |F(x) - F(c)| < \varepsilon.$$

To prove part 2, that F is differentiable with $F'(c) = f(c)$ for every $c \in [a, b]$, we need to prove that these quantities fit appropriately into the definition of differentiability (see Section 8.4). This means proving that for every $c \in (a, b)$ it is true that

$$\lim_{x \to c} \frac{F(x) - F(c)}{x - c} = f(c).$$

Translating further using the definition of limit (see Section 7.10), this means we need to prove that for every $c \in (a, b)$ it is true that

$$\forall \varepsilon > 0 \, \exists \delta > 0 \text{ such that if } 0 < |x - c| < \delta \text{ then } \left| \frac{F(x) - F(c)}{x - c} - f(c) \right| < \varepsilon.$$

Proving this last statement isn't as hard as it looks—the last expression can be simplified by thinking about the meaning of $F(x) - F(c)$ and using some clever algebra. There do remain subtleties to deal with, and these will likely be considered in your course before you see a proof. But the work we've just done in sorting out what is required should make any proof you see more comprehensible.

9.9 Looking ahead

As usual, an Analysis course will cover this chapter's material in detail and will fill in a lot of the gaps. As noted in Section 9.4, it might use variants of the definitions, but the use of approximations will almost certainly be similar. The introduction of the definitions might be different, however. Here, I have treated functions and graphs in the abstract, but your lecturer might encourage you to work first in an applied context, perhaps approximating distance travelled based on information about the way speed changes with time, or approximating the energy required to stretch a spring based on information about the way the force exerted changes as it is stretched. You might like to think now about how those concepts would relate to the abstract ideas presented in this chapter.

Later courses will extend the idea of integrability. Some will extend it to other domains. In this chapter all the domains were closed-interval subsets of the real numbers; a course on multivariable calculus might consider integrals for functions of more variables. A function $f : \mathbb{R}^2 \to \mathbb{R}$, for instance, takes points of the form (x, y) as inputs and returns real numbers as outputs, meaning that its graph can be thought of as a surface in three dimensions. In this context the domain for an integral is a subset of the plane, and the integral can be thought of as the volume under a surface. How do you think upper and lower sums would be calculated?

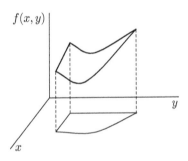

Extending in another direction, it is also possible to define functions $f : \mathbb{C} \to \mathbb{C}$, which opens up more possibilities for integration. In a course on complex analysis you will learn about integrating along curves in the complex plane, which naturally introduces more variation because there are many different curves joining any two points. Complex analysis is extremely elegant and it turns out that for important classes of functions the integral is the same no matter what path is taken, meaning that integrals around curves that start and finish at the same point must be zero. And for some functions it is possible to evaluate integrals simply by calculating function values at certain points. These results lead to applications for real-valued functions: some real integrals that are ordinarily hard to calculate can be found by integrating around a semicircle in the complex plane whose straight edge is part of the real line, then taking limits to make the edge into the whole of the real line and inferring the integral along the real line from the complex integral.

Finally, as mentioned in Section 9.6, you might learn about different types of integrability. In a course on measure theory, for instance, you might learn about Lebesgue[9] integrability. Some functions that are not Riemann integrable are Lebesgue integrable, including this function from Section 9.5:

$$f : \mathbb{R} \to \mathbb{R} \text{ given by } f(x) = \begin{cases} 1 \text{ if } x \in \mathbb{Q} \\ 0 \text{ if } x \notin \mathbb{Q} \end{cases}.$$

We showed that this is not Riemann integrable on any interval $[a, b]$. But it *is* Lebesgue integrable, and its Lebesgue integral is zero for reasons associated with the distribution of the rational numbers on the number line. Such topics are more advanced, but we will do some initial work on rational numbers in Chapter 10.

---

[9] This is a French name so it is pronounced, roughly, 'Le-bayg'.

CHAPTER 10

The Real Numbers

This chapter introduces rational and irrational numbers and their relationships with decimal expansions. It discusses axioms for the real numbers, and explains how the completeness axiom is necessary to distinguish the reals from the rationals and to prove intuitively compelling results about sequences and functions.

10.1 Things you don't know about numbers

You know quite a bit about numbers. And you might think that advanced mathematics is not really about numbers at all, but about general formulas and abstract relationships. To some extent, that is true. However, most undergraduate mathematics degrees involve extended study of the theory of numbers. Some might be in a course called Number Theory, which—at least to begin with—is usually about divisibility properties of integers. Did you know, for instance, that an integer is divisible by 3 if and only if the sum of its digits is divisible by 3? If you did know that, do you know why it is true? This kind of thing is cute but it is not a pointless curiosity; it occurs due to fundamental properties of the base 10 number system. Learning about such phenomena will extend your knowledge 'upward' in the sense discussed in Section 3.3. Analysis, as usual, is about the 'downward' direction—about the theory underlying the mathematics of real numbers.

As a taster, here is a number represented in two ways:

$$1/7 = 0.142857142857142857\ldots$$

Notice that the decimal representation is *repeating* (or *recurring*). Is this a coincidence? What other numbers have this property?

Everyone thinks of decimals as a natural way to represent numbers, not least because they appear that way on a calculator screen. It's good to know how to work your calculator (I do not say that lightly—lots of errors are made when people don't), but you won't need it in Analysis. In fact, I've been known to take calculators away when students reach for them in class. This is partly because smart mental arithmetic is often quicker. And it is partly because of what it means to have a mathematically mature view of numbers. For instance, I recently set an exam in which the answer to one question was $\frac{1}{6}(e-1)$. Many students found this but then got out a calculator and gave a final answer like 0.28638030. Those with more nous,[1] however, left the answer as $\frac{1}{6}(e-1)$. This is what a mathematician would do, because $\frac{1}{6}(e-1)$ is a perfectly good number. Indeed, the decimal version is less accurate no matter how many decimal places are given.

But mostly I remove calculators because Analysis is not really about individual numbers as answers; it is about the structures behind the numbers. A calculator, by necessity, obscures those structures—you get an answer but you don't see why it is correct. Hitting the buttons '1 ÷ 7 =' on a calculator will return the first eight or ten digits, which might not be enough to show up the repeating pattern. Using a computer algebra system will give a lot more digits and might highlight a pattern, but it won't explain why that pattern occurs. In advanced mathematics, it's the *why* that we're interested in, and in this case an explanation is pretty accessible.

10.2 Decimal expansions and rational numbers

It is not a coincidence that the number 1/7 has a repeating decimal expansion. This occurs because 1/7 is a *rational number*. The set of all rational numbers is denoted by $\mathbb{Q}$, and here is the appropriate definition:

Definition: $x \in \mathbb{Q}$ if and only if $\exists p, q \in \mathbb{Z}$ (with $q \neq 0$) such that $x = p/q$.

---

[1] This is a British-English word. It is pronounced like 'house', and my computer's dictionary defines it as 'common sense; practical intelligence'.

In informal terms, x is rational if and only if it can be written as a 'fraction'. It's fine to think of it that way, but be careful with the language because people tend to think of fractions as 'small' and the definition doesn't specify that. For instance, 32800/7 is a perfectly good rational number. It has a repeating expansion too, look:

$$32800/7 = 4685.714285714285714285\ldots$$

In fact, the expansion not only repeats, it repeats with the same *period* as the expansion of 1/7: every six digits. Indeed, the repeating digits are the same ones in the same order. If that were a coincidence it would be pretty weird. So why does it happen?

Doing long division answers this question. I'm not sure what schoolteachers say when describing long division these days, but I learned this inelegant but brief phrasing:

Seven into one doesn't go.	$0.$ $7\overline{)1.0\ 0\ 0\ 0\ 0\ 0}$
Seven into ten goes one remainder three.	0.1 $7\overline{)1.0\ ^30\ 0\ 0\ 0\ 0}$
Seven into thirty goes four remainder two.	$0.1\ 4$ $7\overline{)1.0\ ^30\ ^20\ 0\ 0\ 0}$
Seven into twenty goes two remainder six.	$0.1\ 4\ 2$ $7\overline{)1.0\ ^30\ ^20\ ^60\ 0\ 0}$
Seven into sixty goes eight remainder four.	$0.1\ 4\ 2\ 8$ $7\overline{)1.0\ ^30\ ^20\ ^60\ ^40\ 0\ 0}$
Seven into forty goes five remainder five.	$0.1\ 4\ 2\ 8\ 5$ $7\overline{)1.0\ ^30\ ^20\ ^60\ ^40\ ^50\ 0}$
Seven into fifty goes seven remainder one.	$0.1\ 4\ 2\ 8\ 5\ 7$ $7\overline{)1.0\ ^30\ ^20\ ^60\ ^40\ ^50\ ^10}$

The pattern repeats at this stage because the division process begins to cycle through the same set of remainders. Indeed, this *must* happen because, when dividing by 7, there are only six possible nonzero remainders. So at most six remainders can come up before the digits start repeating.

And this observation is straightforward to generalize: when dividing by $q \in \mathbb{N}$ there are at most $q - 1$ possible nonzero remainders, so the digits must repeat with period at most $q - 1$.

This does not mean that the period has to be exactly $q - 1$. For instance,

$$8/11 = 0.72727272\ldots \quad \text{and} \quad 2/3 = 0.66666666\ldots.$$

And some rational numbers lead to zero remainders beyond some point. For instance,

$$7/8 = 0.8750000\ldots, \text{which we write as } 7/8 = 0.875.$$

But it does mean that every rational number has a repeating or terminating decimal expansion. I think that's a nontrivial thing to know and it's nice that the explanation is so simple. We can take it further, though, by asking the question that has come up repeatedly in this book: is the converse true? Does every repeating decimal expansion represent a rational number?

The answer to this is 'yes' as well. Probably it's easiest to see why by working with a specific number and applying an argument similar to those used for geometric series in Section 6.4.

Let $\qquad x = \qquad 57.257257257257\ldots.$

Then $\qquad 1000x = 57257.257257257257\ldots.$

So $\qquad 1000x - x = 57200,$

i.e. $\qquad 999x = 57200.$

So $\qquad x = \dfrac{57200}{999}.$

This argument could be adapted to deal with any[2] repeating decimal expansion (how?). So rational numbers are precisely those with repeating decimal expansions. That's even more nontrivial—it says something fundamental about the relationships between properties of numbers and their representations. I think it's a shame that people are not taught

---

[2] The potential problems discussed in Section 6.1 do not cause trouble in this case. To see why, refer to Section 6.3 and note that $x = \dfrac{572}{10}\left(1 + \dfrac{1}{10^3} + \dfrac{1}{10^6} + \ldots\right)$.

about it earlier, as the mathematics needed to understand it is pretty straightforward. But you see what I mean about there being plenty left to learn.

While on the subject of decimals, we should sort out something that every undergraduate mathematics student ought to know. Here it is:

$$0.99999999\ldots = 1.$$

This tends to upset people. Their intuition tells them that $0.99999999\ldots$ is a tiny bit less than 1, because they imagine writing down the number, so that the 9s get added in a process that 'never ends' and the written number 'never gets to 1'. Of course, that's perfectly reasonable: a number like 0.99999999 is indeed slightly smaller than 1. But it has only finitely many digits. When mathematicians write '$0.99999999\ldots$' or '$0.\dot{9}$', they do not imagine the process of writing down the 9s. Those symbols mean that the 9s, all infinitely many of them, are already there. So what is the difference between $0.99999999\ldots$ and 1? It has to be zero, meaning that the numbers are equal.

The initial intuition is persistent, though, so here are a couple more ways to defeat it. Those who like algebra might like this:

Let $x = 0.99999999\ldots.$

Then $10x = 9.99999999\ldots.$

So $(10 - 1)x = 9,$

i.e. $9x = 9.$

So $x = \dfrac{9}{9} = 1.$

Or how about even simpler arithmetic? Everyone believes that

$$1/3 = 0.33333333\ldots.$$

Now just multiply both sides by 3.

These are not tricks. It is just that much intuition is based on experience with finite objects, which means that things go awry when people begin thinking about infinite ones—infinite decimal expansions, in this case. In fact, these ideas can be related to limits of infinite sequences, because a decimal expansion can be thought of as the limit of a sequence in which each term contains an extra digit: the sequence $0.9, 0.99, 0.999, 0.9999, \ldots$, for instance, has limit 1. Analysis courses vary in the links they explore between sequences and real numbers, but you might well see these ideas extended.

10.3 Rational and irrational numbers

Here we will move on to contrasts between rational and *irrational* numbers. There are lots of rational numbers, which raises the question of whether it is possible to write every number in the form p/q. There are, after all, an awful lot of combinations of p and q.

Think again about decimal expansions, though, and the picture starts to look different. All rational numbers can be represented as repeating decimals, and clearly there are many decimal expansions that do not repeat. It's easy to imagine taking a single repeating decimal expansion and 'messing it up' in numerous ways to get non-repeating ones (easy but not trivial—we'd need to mess it up enough). The decimal expansions idea provides insight, therefore, but it makes irrationals quite hard to get hold of: to express one fully we'd have to write down infinitely many digits, and no one can do that.

But it is not too hard to establish that some familiar numbers are irrational by using indirect methods. Roughly speaking, the word 'indirect' is used when instead of proving (directly) that something is true, we prove (indirectly) that something *can't not* be true. That sounds inelegant but in fact it can lead to some rather beautiful proofs. One classic is a proof by contradiction that $\sqrt{2}$ is irrational, a version of which is shown here. To understand this proof, you will need to know that the notation $2|p$ is read aloud as '2 divides p', meaning that 2 is a factor of p. Notice also that it's important in proofs like this to make sure that the symbols '|' and '/' are distinguishable.

Claim: $\sqrt{2}$ is irrational.

Proof: Suppose for contradiction that $\sqrt{2} \in \mathbb{Q}$.

Then $\exists \, p, q \in \mathbb{Z}$ (with $q \neq 0$) such that $\sqrt{2} = p/q$ and p and q have no common factors.

This implies that $2 = p^2/q^2$ so $2q^2 = p^2$.

Hence $2 | p^2$.

But then $2 | p$ because 2 is prime.

Say $p = 2k$ where $k \in \mathbb{Z}$.

Then $2q^2 = 4k^2$, so $q^2 = 2k^2$.

So $2 | q^2$.

But then $2 | q$ because 2 is prime.

So p and q have common factor 2.

But this gives a contradiction.

Hence $\sqrt{2} \notin \mathbb{Q}$.

For what other numbers would such a proof work? Could we replace $\sqrt{2}$ with $\sqrt{3}$ and still have a valid argument? Clearly we couldn't replace it with $\sqrt{4}$, because $\sqrt{4}$ isn't irrational. But at what step would the proof break down? Are there multiple steps that don't apply for $\sqrt{4}$, or is there just one key step that is not valid? And could we replace $\sqrt{2}$ with $\sqrt{6}$? If so, what else would have to change? As usual, you should get in the habit of asking yourself such questions. And note that by no means all irrational numbers arise as square roots. In fact, in an important sense, there are many 'more' irrationals than rationals—look out for a proof of this.

As usual, a typical Analysis course will introduce rational and irrational numbers then do some work on how they combine. For instance, multiplying together two rational numbers always gives another rational. Why, exactly? And does multiplying together two irrationals always give an irrational? Be careful here—the answer is 'no', and lecturers like to ask questions like this to make sure that students are thinking carefully. What about multiplying a rational by an irrational? This, again, can trip people up, because zero is rational and multiplying any number by zero gives zero. But, if the rational number is not zero, we get another irrational. This can also be proved by contradiction, as follows.

Theorem: If $x \in \mathbb{Q}$, $x \neq 0$ and $y \notin \mathbb{Q}$, then $xy \notin \mathbb{Q}$.

    Proof: Let $x \in \mathbb{Q}$ and $x \neq 0$.

Then $\exists\, p, q \in \mathbb{Z}$ (with $q \neq 0$) such that $x = p/q$, and $p \neq 0$ because $x \neq 0$.

Let $y \notin \mathbb{Q}$ and suppose for contradiction that $xy \in \mathbb{Q}$.

This means that $\exists\, r, s \in \mathbb{Z}$ (with $s \neq 0$) such that $xy = r/s$.

But then $y = \dfrac{q}{p} \times \dfrac{r}{s} = \dfrac{qr}{ps}$.

Now $qr \in \mathbb{Z}$ and $ps \in \mathbb{Z}$ because $p, q, r, s \in \mathbb{Z}$.

Also $ps \neq 0$ because $p \neq 0$ and $s \neq 0$.

So $y \in \mathbb{Q}$.

But this contradicts the theorem premise.

Hence $xy \notin \mathbb{Q}$.

Proofs by contradiction pop up a lot in work with irrational numbers, precisely because it is hard to work with irrationals directly. Effectively the thinking goes, 'I know this number is going to be irrational, but rationals are easier to work with so let's suppose it's rational and show that something goes wrong'. This is exactly how proof by contradiction works.

10.4 Axioms for the real numbers

We have established that some real numbers are rational and some are irrational. But lots of things are true for all real numbers, and these *axioms* are the subject of this section.

Recall that in Section 2.2 I listed these axioms:

$\forall a, b \in \mathbb{R}, a + b = b + a$ [commutativity of addition];
$\exists\, 0 \in \mathbb{R}$ s.t. $\forall a \in \mathbb{R}, a + 0 = a = 0 + a$ [existence of an additive identity].

You no doubt believe that these axioms are true. So does everyone else. But how do we know? The philosophically interesting answer is that *we don't*. It's not like anyone has checked every possible pair of real numbers a and b to make sure it really is always true that $a + b = b + a$. Philosophically, Platonists believe that the real numbers are out there and that an axiom like this is a human attempt to capture one of their properties. Formalists believe that an axiom like this is a definition stipulating a property of a set that we can choose to call the real numbers; for a formalist, $2 + 3 = 3 + 2$ is true *because the axiom says so*. This is not a problem and, depending on the structure of your degree, you might do a course that constructs sets that satisfy the expected axioms for the natural numbers, the integers, the rationals and the reals. That would be too detailed for a book like this, but it's good to begin thinking about the philosophical assumptions behind even simple mathematics.

In any case, those are just two axioms, and the real numbers satisfy a whole lot more; below you can find a list. Some of these axioms have names, and these are also listed. Which name do you think goes with which axiom? (This isn't an unreasonable question—given what you already know, you'll be able to get most of them right.)

Axioms

1. $\forall a, b \in \mathbb{R}, a + b \in \mathbb{R}$.
2. $\forall a, b \in \mathbb{R}, ab \in \mathbb{R}$.
3. $\forall a, b, c \in \mathbb{R}, (a + b) + c = a + (b + c)$.
4. $\forall a, b \in \mathbb{R}, a + b = b + a$.
5. $\exists\, 0 \in \mathbb{R}$ s.t. $\forall a \in \mathbb{R}, a + 0 = a = 0 + a$.
6. $\forall a \in \mathbb{R} \; \exists\, (-a) \in \mathbb{R}$ s.t. $a + (-a) = 0 = (-a) + a$.
7. $\forall a, b, c \in \mathbb{R}, (ab)c = a(bc)$.
8. $\forall a, b \in \mathbb{R}, ab = ba$.
9. $\exists\, 1 \in \mathbb{R}$ s.t. $\forall a \in \mathbb{R}, a \cdot 1 = a = 1 \cdot a$.
10. $\forall a \in \mathbb{R} \backslash \{0\} \; \exists\, a^{-1} \in \mathbb{R}$ s.t. $aa^{-1} = 1 = a^{-1}a$.
11. $\forall a, b, c \in \mathbb{R}, a(b + c) = ab + ac$.
12. $\forall a, b \in \mathbb{R}$, exactly one of $a < b$, $a = b$ and $a > b$ is true.
13. $\forall a, b, c \in \mathbb{R}$, if $a < b$ and $b < c$ then $a < c$.
14. $\forall a, b, c \in \mathbb{R}$, if $a < b$ then $a + c < b + c$.
15. $\forall a, b, c \in \mathbb{R}$, if $a < b$ and $c > 0$ then $ca < cb$.

Axiom names

closure under multiplication	associativity of multiplication
existence of a multiplicative identity	trichotomy
associativity of addition	commutativity of addition
existence of multiplicative inverses	closure under addition
commutativity of multiplication	existence of additive inverses
transitivity	distributivity of multiplication over addition
existence of an additive identity	

The axiom names are a bit long and students often don't learn them. In one sense that doesn't matter—you can use an axiom without knowing its name. But names are useful both for identifying links across subjects and for effective communication. For instance, both addition and multiplication are commutative—they share this property so it's useful to have a word to describe it. And mathematicians also work with complex numbers, functions, matrices, symmetries, vectors, and so on—many of these objects can be added or multiplied, and we can ask whether addition and multiplication remain commutative. Moreover, restricted sets of axioms define structures such as *vector spaces*, *groups*, *rings* and *fields*, which are studied in work on linear algebra and abstract algebra. Naming the axioms makes it easier to compare and communicate about these structures.

Returning to the real numbers, though, here is a question. In which of the axioms could we replace $\mathbb{R}$ with $\mathbb{Q}$? Look back and decide.

10.5 Completeness

The answer to the preceding question is 'all of them': all fifteen axioms still apply if we replace $\mathbb{R}$ with $\mathbb{Q}$. Make sure you believe this. So that long list of axioms is not sufficient to distinguish the real numbers from the rationals. We need something else, and that something is known as *completeness*.

Completeness is not a complicated idea, but to understand it you need to understand the idea of a *supremum* of a set $X \subseteq \mathbb{R}$.

Definition: U is the *supremum* of $X \subseteq \mathbb{R}$ if and only if

1. $\forall x \in X, x \leq U$;
2. if u is any upper bound for X, then $U \leq u$.

The supremum is sometimes referred to as the *least upper bound*. Can you see why? Point 1 in the definition means that U is an upper bound for X (see Section 2.6), and point 2 means that it is the least of all the possible upper bounds. Students often go further in the direction of informal thinking and assume that the supremum of a set is its maximum or largest element. Unfortunately, that is not correct, because not every set has a maximum element. Some sets do: the set $[1, 5] = \{x \in \mathbb{R} | 1 \leq x \leq 5\}$ has maximum 5, and 5 is also its supremum (check against the definition). But the set $(1, 5) = \{x \in \mathbb{R} | 1 < x < 5\}$ does not have a maximum element: whatever $x \in (1, 5)$ we pick, there will be a bigger one. The set $(1, 5)$ still has a supremum, though, and its supremum is also 5 (again, check). It just happens that 5 is not in $(1, 5)$. So it is important to pay attention to the definition and to avoid being swayed by the related informal idea. Students who don't avoid being swayed often find it difficult to construct proofs involving suprema,[3] not because the definition is logically complicated but because they think they understand the concept so they don't think to invoke it.

A similar comment applies to the definition of the *infimum* of a set, which is also known as the *greatest lower bound*. Can you construct

[3] As noted in Section 9.7, *supremum* is the singular, *suprema* is the plural—this is like *maximum* and *maxima*.

the definition of *infimum*? And why should we avoid assuming that the infimum of a set is its minimum element?

With the definition of supremum in place, we can introduce completeness:

Completeness Axiom: Every nonempty subset of $\mathbb{R}$ that is bounded above has a supremum in $\mathbb{R}$.

The completeness axiom captures the distinction between the reals and the rationals; replacing $\mathbb{R}$ with $\mathbb{Q}$ in this axiom gives a statement that *isn't* true. For instance, the set $\{x \in \mathbb{Q} \mid x^2 < 2\}$ does not have a supremum in $\mathbb{Q}$; its supremum is $\sqrt{2}$, which is in $\mathbb{R}$ but not in $\mathbb{Q}$—if we lived in a world with only rational numbers, and we zoomed in on a number line, we'd find a gap where $\sqrt{2}$ ought to be. Because of this, people sometimes describe the completeness axiom informally by saying

'there are no holes in the number line.'

This doesn't surprise anyone because everyone has always assumed that there are no holes in the number line. But, again, Analysis highlights the philosophical assumptions we all make without thinking about them. We want to assume that there are no holes in the number line, so to axiomatize the number system properly we need to state that explicitly.

Focusing on completeness permits a deeper understanding of results from other Analysis topics. For instance, remember this potential theorem from Section 5.4?

- Every bounded monotonic sequence is convergent.

This one is true, which for most people seems intuitively reasonable: if a sequence (a_n) is increasing, say, and bounded above by u, then infinitely many terms must 'fit in' between a_1 and u. In fact, the limit will be the supremum U of the set of all sequence terms $\{a_n \mid n \in \mathbb{N}\}$. Note that the sequence might or might not have terms equal to its limit, and correspondingly U might or might not be in $\{a_n \mid n \in \mathbb{N}\}$.

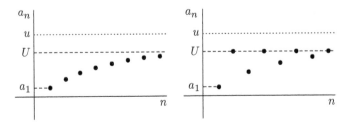

Either way, this theorem is true only because $\mathbb{R}$ is complete. If it were not, some apparently convergent sequences would not have limits. Consider, for example, the sequence in which the nth term is the n-decimal-place approximation to $\sqrt{3}$: the sequence $1.7, 1.73, 1.732, \ldots$. If we lived in a world with only rational numbers, this sequence would exist (every term is rational—for example $1.732 = 1732/1000$) but its limit would not.

Similarly, consider this theorem from Section 7.9:

Intermediate Value Theorem:
Suppose that f is continuous on $[a, b]$ and that y is between $f(a)$ and $f(b)$. Then $\exists\, c \in (a, b)$ such that $f(c) = y$.

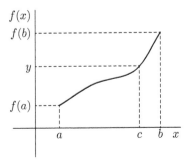

This is also true, and we need completeness to prove it. If the reals were not complete, then function graphs would have 'holes' and there might be a value y with no appropriate c. A typical proof involves considering the set $X = \{x \in [a, b] | f(x) < y\}$; this is a bounded subset of $\mathbb{R}$ so it must have a supremum c in $\mathbb{R}$ by the completeness axiom, and it must be true that $f(c) = y$. The details need fleshing out, but thinking about the diagrams below might help you to understand a proof—where would $c = \sup X$ be in each case?

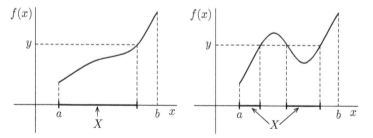

10.6 Looking ahead

This chapter gives an introduction to ideas about the real numbers that you might encounter in Analysis. Depending on your degree programme, you might not go much beyond this—if you work mostly on applied mathematics then your studies will rely on properties of numbers but the axioms and definitions will remain in the background. If you do more pure mathematics, however, these ideas will be extended in a number of directions.

One thing you might study is the classification of numbers according to their status as solutions to different types of equation. Rational numbers, for instance, are 'nice' in the sense that they are solutions to linear equations like $3x - 4 = 0$. Irrationals are less nice in that they are not, but some are solutions to quadratic equations like $x^2 - 2 = 0$. In general, a number is said to be *algebraic* if it is a solution to an equation of the form $a_n x^n + \ldots + a_2 x^2 + a_1 x + a_0 = 0$ with integer coefficients. Some irrationals, though, are not even that nice. For instance, e and π are both *transcendental* numbers, meaning that they do not satisfy any such equation; you might see this proved. And you might study properties of the sets of solutions to algebraic equations—the subject known as Galois theory is about the abstract group structures formed by such solutions and it has far-reaching implications in geometry as well as in abstract algebra.

Another thing you might learn more about is axiomatic systems. As I mentioned, structures such as vector spaces, groups, rings and fields are all defined by subsets of the axioms listed in Section 10.4; most mathematics students study both specific examples of these structures and general theorems about their properties. For now, you might like to think about which familiar mathematical sets—$\mathbb{N}$, $\mathbb{Z}$, $\mathbb{Q}$, $\mathbb{C}$ (the set of all complex numbers), the set of all three-component vectors, the set of all 2×2 matrices, and so on—satisfy which ones. Notice that some axioms do not hold in certain structures—$\mathbb{Z}$, for instance, has additive inverses but not multiplicative inverses. And some axioms might not apply at all: the order axioms (the ones about inequalities) make no sense for complex numbers or for matrices—what would it mean to say that one matrix is 'less than' another? This means that there are fundamental differences in the types of mathematics that can be done in these systems and in the theorems and theories that apply.

Finally, you might take a course that does foundational work. This might characterize the rational numbers in terms of equivalence relations, clarifying what it means to say that $\frac{1}{2} = \frac{2}{4} = \frac{3}{6} = \ldots$ and proving that all the algebraic properties of rationals respect these relationships. Or it might go further back and use set theory to construct the naturals, integers, rationals and reals. Such courses require students to stop taking even their basic mathematical knowledge for granted, so they are challenging but they get to the heart of what mathematics is about.

Conclusion

This short concluding chapter reviews ideas from the book and suggests things to bear in mind when studying Analysis.

This book hits all the main topics in Analysis—sequences, series, continuity, differentiability, integrability and the real numbers—and explains the main definitions in detail. But it doesn't try to do the same for all the theorems and proofs that might be included in textbooks or in a taught Analysis course. Rather, it aims to teach some useful skills for learning from such a presentation. With that in mind, you might want to review Part 1 before you put the book down—now that you know more about the content, you might have a more informed perspective on the general suggestions.

Because of the book's aims, I have been able to indulge myself by selectively including things that I find particularly interesting. Some of these are elegant arguments, or good tricks, or graphical representations that provide intuitive insight. Some are (to me) counterintuitive and therefore fascinating: the study of series, in particular, throws up conceptual surprises that give many students a sense that although they know a lot of mathematics, there remain some deep ideas to understand.

Indeed, for many people, understanding Analysis means not only extending their existing knowledge but also gaining a new perspective by examining its underlying assumptions. I sometimes wonder whether students will be impatient with this—whether they will want to learn higher-level stuff rather than digging around in the foundations. But in fact students often tell me that they find it really interesting to learn about the ideas behind long-familiar mathematics. Even if Analysis turns out not to be your thing—if you find that you prefer working on real-world modelling problems, for instance—I hope this book gives you a sense of what Analysis achieves and why people think it's an important thing to study.

In any case, this is the book's conclusion but it is just the beginning of your study of Analysis. A small number of readers will spend a lifetime engaged in work on advanced versions of these ideas, and many more will take numerous courses extending them into topics like complex analysis, differential geometry, metric spaces, and topology. All will take at least one or two courses that include the material presented here and that fill in the gaps. With that in mind, I'll leave you with a quick review of some important things to remember.

First, every student of Analysis should pay attention to the definitions. Advanced mathematics relies heavily on its definitions and you should look for them everywhere, especially in new topics—you have little hope of understanding what is going on if you don't. Moreover, you need the definitions to come to mind. About once a week I have a conversation with my tutees[1] that goes like this:

> STUDENT: I just don't get what's going on with concept X.
>
> ME: Okay. What does concept X mean?
>
> STUDENT: I don't really know—I know it's something to do with . . . um . . . well . . . I can't explain it.
>
> ME: Alright, that's fine. What do we do when we don't know what something means in advanced mathematics?
>
> STUDENT: (looking a bit embarrassed because I ask this every week, and shuffling notes to find the appropriate page) We look at the definition . . .

This is not because my tutees are no good. On the contrary, they're pretty sharp and hardworking. But they're new to advanced mathematics and they're accustomed to learning procedures rather than thinking about concepts and logical arguments. Sometimes it just doesn't occur to them to treat definitions as the source of meaning, so they experience

---

[1] In in the UK I have 'tutees'—'advisees' is a rough US equivalent.

generalized confusion about a whole topic without realizing that they could sort it out by going back to the place where the key concept was introduced. Remember to do that and your life will be easier.

Second, when reading lecture notes or a textbook, read properly. Read whole mathematical sentences, not just the algebraic bits. Ideally, read out loud, especially if you're stuck. Another regular conversation I have with students, often in our Mathematics Learning Support Centre, goes like this:

STUDENT: (looking a bit nervous) I don't understand this proof.

ME: Okay, that's fine. Do you want to start reading it out loud so we can see where you get stuck?

STUDENT: Um, okay. 'Let f be a function from $\mathbb{R}$ to $\mathbb{R}$ that is differentiable . . .' [keeps reading until about halfway through the proof] . . . Oh! I get it now.

ME: Jolly good. Do you want me to say any more about it?

STUDENT: No that's alright. Thank you very much.

This is great for me because I get thanked when I haven't actually done anything. And of course it doesn't happen every time—sometimes we do bump into a problem and I help the student sort it out. But it's cheering to see how many students can resolve their own difficulties just by listening to their own voices say the words. Usually I point out that they didn't need me at all, and they go away with improved confidence as well as improved understanding.

On a related point, do take opportunities to speak about mathematics. You will be terrible at this at first—ideas that seem fairly clear in your head will come out of your mouth as a garbled jumble with dodgy logic and worse grammar. But that doesn't matter. If you're going to write mathematics fluently in exams and project reports and so on, you need to own it, and you won't get to that point without practice. I try to facilitate this by giving students lots of time to talk about mathematics in lectures, but not everyone does that so you might want to set up a study group to arrange such opportunities for yourself. At any rate, you'll get through the ropey phase faster if you just embrace it.

Third, look for links between diagrams or other informal representations and the formal mathematics. I say this because Analysis students often tell me that they 'get the ideas' but don't understand much of their notes. Usually these people are telling the truth—when asked to explain their understanding, they can do so by gesturing or by drawing diagrams, and it's clear that their thinking is fairly accurate. What they haven't done is linked that understanding in detail to the expressions that appear in a relevant definition, theorem or proof. Usually this can be fixed by finding the appropriate notes and going through them slowly, explicitly linking each phrase to a diagram drawn on paper and adding labels where possible. This process sometimes makes people a bit nervous because they think they should be able to read fast. I don't have much to say about that except that it isn't true. Slowing down and understanding such links firms up your grasp on both the meanings of the symbolic sentences and the subtleties of the diagrams, thus rendering everything more memorable and allowing you greater flexibility in translating between the two. After doing it a few times you'll be able to speed up again, but you'll find that you no longer particularly want to.

Fourth, when trying to write mathematics, get something down on the page. Students sometimes seem paralyzed in the face of a blank page—they are unwilling to write anything down because they don't know how to produce a full and perfect calculation or argument. Lecturers have a lot to answer for here. We walk into lectures and give pristine presentations, writing down fully correct proofs with no hesitation. This gives at least some students the impression that they should be able to do that too. But the reason I can give a polished presentation is that I *prepare*. I dare say that you too would prepare if you were going to stand up and give a 50-minute lecture. I don't want to end up looking like an idiot in front of 200 people, so I make sure I know in advance what I'm going to say and how I'm going to say it. If you saw me genuinely working on a piece of mathematics—even trying to draft a clear model answer to a problem in my own Analysis course—you'd see me making notes and drawing diagrams and doing bits of algebra and re-writing sections in which I've decided it would be better to say something in a different order. I'm probably a bit faster than you are, but my thinking processes are much the same.

A fancy-sounding phrase to describe writing ideas down is 'outsourcing your cognition to the environment'. In essence, you get the paper and ink to do some of the work for you by storing your ideas in a form that allows you to reflect on them and spot new links. This advice applies particularly when getting started on a proof. Write down what you know (the premises) in terms of the relevant definitions, then write down what you want to prove (the conclusion) in terms of the relevant definitions, then look at them both and think. If that doesn't get you started, write down some theorems that might conceivably be useful, or draw a diagram, or try something out using a specific example. There is no way you will be able to hold all the possibly relevant things in mind at once, so don't waste time trying—get them on the page. And don't worry too much about writing your ideas 'mathematically'. Students often think they should write everything in symbols, but I usually advise them to express their idea in whatever way they can, then to convert it once they've got something written down. In reality, though, this is often a red herring. A mathematician won't see much difference between a correct wordy expression of an idea and a correct symbolic one. Certainly I go out of my way when marking exams to give credit for any clear communication of appropriate understanding.

Fifth, in your studies in general, aim for a balance between persevering and taking proper breaks. You'll never get anywhere if you give up as soon as something gets a bit difficult, but there are no prizes for the student who can sit at a desk the longest. If you keep going when you're tired you'll become less and less effective; if you do that repeatedly you'll burn yourself out. I was reminded of this recently. It was week ten of our semester and I looked at my tutees and they looked exhausted; then I went into a lecture and the whole lecture class looked exhausted too. To be honest I was pretty shattered myself—I think we'd all hit the wall at more or less the same time. There's nothing wrong with you if you're worn out after ten weeks of trying every day to learn new, difficult mathematics— you'd never criticize anyone else for that. So take a break if you need one, and try using it to do something practical. Go to the gym, or go out shopping, or clear out your wardrobe, or cook a big dinner for your friends. Doing physical things will often clear your brain out nicely.

Sixth, embrace opportunities to be wrong. You'll get a lot of these in an Analysis course—Analysis is a gift to a lecturer in that it contains a lot of true theorems with plausible but untrue converses. This provides many opportunities to ask challenging true/false or multiple-choice questions, both on tests and in lectures. I was lucky this year to have a lecture class who were willing to vote on such questions by a show of hands, even though it quickly became clear that everyone would be wrong a lot of the time. I always reminded the students that I didn't care whether they were right or wrong—I just cared that they were thinking about the question and were willing to change their minds in response to a good reason—but in any case I admired their willingness to engage. And I was delighted with the atmosphere that took over the room when we did this. People were properly invested in their answers but everyone seemed to remain good-humoured when they were wrong too.

Indeed, I think good humour is probably key to a good undergraduate learning experience. My favourite thing about my current tutees, which makes me think they'll all do well in life as well as in their studies, is this. They often make errors. But usually, when someone realizes they've made an error, they start laughing. Then everyone else starts laughing too. Then we sort out whatever it is, and move on. I think this is mostly a personality thing—I didn't do anything to start it off—but I really admire it. People often feel a bit insecure when they're taking on a new challenge, and when you feel insecure it's easy to think that being wrong about something reflects badly on you. But it really doesn't. Everyone is wrong a lot, and confused a lot, when learning a subject like Analysis. If you can be the person who sets the tone by both laughing about your errors and working to improve your understanding, you'll do your friends, your class, your lecturer and yourself a big favour.

BIBLIOGRAPHY

Aberdein, A. (2005). The uses of argument in mathematics. *Argumentation,* *19,* 287–301.

Ainsworth, S. (2008). The educational value of multiple-representations when learning complex scientific concepts. In J. K. Gilbert, M. Reiner, & M. Nakhleh (Eds.), *Visualization: Theory and practice in science education* (pp. 191–208). New York: Springer.

Ainsworth, S., & Burcham, S. (2007). The impact of text coherence on learning by self-explanation. *Learning and Instruction, 17,* 286–303.

Alcock, L. (2010). Mathematicians' perspectives on the teaching and learning of proof. In F. Hitt, D. Holton, & P. W. Thompson (Eds.), *Research in Collegiate Mathematics Education VII* (pp. 63–92). Washington DC: MAA.

Alcock, L. (2013a). *How to study as a mathematics major.* Oxford: Oxford University Press.

Alcock, L. (2013b). *How to study for a mathematics degree.* Oxford: Oxford University Press.

Alcock, L., Attridge, N., Kenny, S., & Inglis, M. (2014). Achievement and be-haviour in undergraduate mathematics: Personality is a better predictor than gender. *Research in Mathematics Education, 16,* 1–17.

Alcock, L., & Inglis, M. (2008). Doctoral students' use of examples in evalu-ating and proving conjectures. *Educational Studies in Mathematics, 69,* 111–29.

Alcock, L., & Inglis, M. (2010). Representation systems and undergradu-ate proof production: A comment on Weber. *Journal of Mathematical Behavior, 28,* 209–11.

Alcock, L., & Simpson, A. (2001). The Warwick analysis project: Practice and theory. In D. Holton (Ed.), *The teaching and learning of mathematics at the undergraduate level* (pp. 99–112). Dordrecht: Kluwer.

Alcock, L., & Simpson, A. (2002). Definitions: Dealing with categories math-ematically. *For the Learning of Mathematics, 22*(2), 28–34.

Alcock, L., & Simpson, A. (2004). Convergence of sequences and series: Interactions between visual reasoning and the learner's beliefs about their own role. *Educational Studies in Mathematics, 57*, 1–32.

Alcock, L., & Simpson, A. (2005). Convergence of sequences and series 2: Interactions between nonvisual reasoning and the learner's beliefs about their own role. *Educational Studies in Mathematics, 58*, 77–100.

Alcock, L., & Simpson, A. (2011). Classification and concept consistency. *Canadian Journal of Science, Mathematics and Technology Education, 11*, 91–106.

Alcock, L., & Weber, K. (2005). Proof validation in real analysis: Inferring and checking warrants. *Journal of Mathematical Behavior, 24*, 125–34.

Alcock, L., & Weber, K. (2010). Referential and syntactic approaches to proving: Case studies from a transition-to-proof course. In F. Hitt, D. Holton, & P. W. Thompson (Eds.), *Research in Collegiate Mathematics Education VII* (pp. 93–114). Washington, DC: MAA.

Almeida, D. (1995). Mathematics undergraduates' perceptions of proof. *Teaching Mathematics and its Applications, 14*, 171–7.

Antonini, S. (2011). Generating examples: Focus on processes. *ZDM: The International Journal on Mathematics Education, 43*, 205–17.

Arcavi, A. (2003). The role of visual representations in the learning of mathematics. *Educational Studies in Mathematics, 52*, 215–41.

Artigue, M. (1991). Analysis. In D. O. Tall (Ed.), *Advanced mathematical thinking* (pp. 167–98). Dordrecht: Kluwer.

Attridge, N., & Inglis, M. (2013). Advanced mathematical study and the development of conditional reasoning skills. *PLoS ONE, 8*, e69399.

Barbé, J., Bosch, M., Espinoza, L., & Gascón, J. (2005). Didactic restrictions on the teacher's practice: The case of limits of functions in Spanish high schools. *Educational Studies in Mathematics, 59*, 235–68.

Bardelle, C., & Ferrari, P. L. (2011). Definitions and examples in elementary calculus: The case of monotonicity of functions. *ZDM: The International Journal on Mathematics Education, 43*, 233–46.

Bergé, A. (2008). The completeness property of the set of real numbers in the transition from calculus to analysis. *Educational Studies in Mathematics, 67*, 217–35.

Bergqvist, E. (2007). Types of reasoning required in university exams in mathematics. *Journal of Mathematical Behavior, 26*, 348–70.

Bergsten, C. (2008). On the influence of theory on research in mathematics education: The case of teaching and learning limits of functions. *ZDM: The International Journal on Mathematics Education, 40*, 189–99.

Bielaczyc, K., Pirolli, P. L., & Brown, A. L. (1995). Training in self-explanation and self-regulation strategies: Investigating the effects of knowledge acquisition activities on problem solving. *Cognition and Instruction, 13*, 221–52.

Biza, I., Christou, C., & Zachariades, T. (2008). Student perspectives on the relationship between a curve and its tangent in the transition from Euclidean geometry to analysis. *Research in Mathematics Education, 10*, 53–70.

Biza, I., & Zachariades, T. (2010). First year mathematics undergraduates' settled images of tangent line. *Journal of Mathematical Behavior, 29*, 218–29.

Bremigan, E. G. (2005). An analysis of diagram modification and construction in students' solutions to applied calculus problems. *Journal for Research in Mathematics Education, 36*, 248–77.

Brown, J. R. (1999). *Philosophy of mathematics: An introduction to the world of proofs and pictures.* New York: Routledge.

Buchbinder, O., & Zaslavsky, O. (2011). Is this a coincidence? The role of examples in fostering a need for proof. *ZDM: The International Journal on Mathematics Education, 43*, 269–81.

Burn, R. P. (1992). *Numbers and functions: Steps into analysis.* Cambridge: Cambridge University Press.

Chater, N., Heit, E., & Oaksford, M. (2005). Reasoning. In K. Lamberts & R. Goldstone (Eds.), *Handbook of cognition* (pp. 297–320). London: Sage.

Chi, M. T. H., Bassok, M., Lewis, M. W., Reimann, P., & Glaser, R. (1989). Self-explanations: How students study and use examples in learning to solve problems. *Cognitive Science, 13*, 145–82.

Chi, M. T. H., de Leeuw, N., Chiu, M.-H., & LaVancher, C. (1994). Eliciting self-explanations improves understanding. *Cognitive Science, 18*, 439–77.

Conradie, J., & Frith, J. (2000). Comprehension tests in mathematics. *Educational Studies in Mathematics, 42*, 225–35.

Copes, L. (1982). The Perry development scheme: A metaphor for learning and teaching mathematics. *For the Learning of Mathematics, 3*(1), 38–44.

Cornu, B. (1991). Limits. In D. O. Tall (Ed.), *Advanced mathematical thinking* (pp. 153–66). Dordrecht: Kluwer.

Cowen, C. (1991). Teaching and testing mathematics reading. *American Mathematical Monthly, 98,* 50–53.

Crawford, K., Gordon, S., Nicholas, J., & Prosser, M. (1994). Conceptions of mathematics and how it is learned: The perspectives of students entering university. *Learning and Instruction, 4,* 331–45.

Crawford, K., Gordon, S., Nicholas, J., & Prosser, M. (1998a). Qualitatively different experiences of learning mathematics at university. *Learning and Instruction, 8,* 455–68.

Crawford, K., Gordon, S., Nicholas, J., & Prosser, M. (1998b). University mathematics students' conceptions of mathematics. *Studies in Higher Education, 23,* 87–94.

Dahlberg, R. P., & Housman, D. L. (1997). Facilitating learning events through example generation. *Educational Studies in Mathematics, 33,* 283–99.

Davis, R. B., & Vinner, S. (1986). The notion of limit: Some seemingly unavoidable misconception stages. *Journal of Mathematical Behavior, 5,* 281–303.

Dawkins, P. C. (2014). How students interpret and enact inquiry-oriented defining practices in undergraduate real analysis. *Journal of Mathematical Behavior, 33,* 88–105.

de Jong, T. (2010). Cognitive load theory, educational research, and instructional design: Some food for thought. *Instructional Science, 38,* 105–34.

de Villiers, M. (1990). The role and function of proof in mathematics. *Pythagoras, 24,* 17–24.

Deloustal-Jorrand, V. (2002). Implication and mathematical reasoning. In A. D. Cockburn & E. Nardi (Eds.), *Proceedings of the 26th International Conference on the Psychology of Mathematics Education* (Vol. 2, pp. 281–8). Norwich, UK: IGPME.

Duah, F., Croft, T., & Inglis, M. (2014). Can peer-assisted learning be effective in undergraduate mathematics? *International Journal of Mathematical Education in Science and Technology, 45,* 552–65.

Dubinsky, E., Elterman, F., & Gong, C. (1988). The student's construction of quantification. *For the Learning of Mathematics, 8*(2), 44–51.

Durkin, K. (2011). *The self-explanation effect when learning mathematics: A meta-analysis.* Evanston, IL: Society for Research on Educational Effectiveness.

Durrand-Guerrier, V. (2003). Which notion of implication is the right one? From logical considerations to a didactic perspective. *Educational Studies in Mathematics, 53*, 5–34.

Edwards, A., & Alcock, L. (2010). How do undergraduate students navigate their example spaces? In *Proceedings of the 32nd conference on research in undergraduate mathematics education*. Raleigh, NC, USA.

Edwards, B. S., & Ward, M. B. (2004). Surprises from mathematics education research: Student (mis)use of mathematical definitions. *American Mathematical Monthly, 111*, 411–24.

Epp, S. (2003). The role of logic in teaching proof. *American Mathematical Monthly, 110*, 886–99.

Even, R. (1993). Subject-matter knowledge and pedagogical content knowledge: Prospective secondary teachers and the function concept. *Journal for Research in Mathematics Education, 24*, 94–116.

Fischbein, E. (1982). Intuition and proof. *For the Learning of Mathematics, 3*(2), 9–18.

Furinghetti, F., Morselli, F., & Antonini, S. (2011). To exist or not to exist: Example generation in real analysis. *ZDM: The International Journal on Mathematics Education, 43*, 219–32.

Giaquinto, M. (2007). *Visual thinking in mathematics*. Oxford: Oxford University Press.

Goulding, M., Hatch, G., & Rodd, M. (2003). Undergraduate mathematics experience: Its significance in secondary mathematics teacher preparation. *Journal of Mathematics Teacher Education, 6*, 361–93.

Güçler, B. (2013). Examining the discourse on the limit concept in a beginning-level calculus classroom. *Educational Studies in Mathematics, 82*, 439–53.

Gueudet, G. (2008). Investigating the secondary–tertiary transition. *Educational Studies in Mathematics, 67*, 237–54.

Hadamard, J. (1945). *The psychology of invention in the mathematical field* (2nd ed.). New York: Dover Publications.

Hardy, N. (2009). Students' perceptions of institutional practices: The case of limits of functions in college level calculus courses. *Educational Studies in Mathematics, 72*, 341–58.

Harel, G., & Sowder, L. (1998). Students' proof schemes: Results from exploratory studies. In A. H. Schoenfeld, J. Kaput, & E. Dubinsky (Eds.),

Research in collegiate mathematics III (pp. 234–82). Providence, RI: American Mathematical Society.

Hazzan, O., & Leron, U. (1996). Students' use and misuse of mathematical theorems: The case of Lagrange's theorem. *For the Learning of Mathematics, 16*(1), 23–6.

Heinze, A. (2010). Mathematicians' individual criteria for accepting theorems and proofs: An empirical approach. In G. Hanna, H. N. Jahnke, & H. Pulte (Eds.), *Explanation and proof in mathematics* (pp. 101–11). New York: Springer.

Hersh, R. (1993). Proving is convincing and explaining. *Educational Studies in Mathematics, 24*(4), 389–99.

Hodds, M., Alcock, L., & Inglis, M. (2014). Self-explanation training improves proof comprehension. *Journal for Research in Mathematics Education, 45*, 62–101.

Housman, D., & Porter, M. (2003). Proof schemes and learning strategies of above-average mathematics students. *Educational Studies in Mathematics, 53*, 139–58.

Hoyles, C., & Küchemann, D. (2002). Students' understanding of logical implication. *Educational Studies in Mathematics, 51*, 193–23.

Iannone, P., Inglis, M., Mejía-Ramos, J., Simpson, A., & Weber, K. (2011). Does generating examples aid proof production? *Educational Studies in Mathematics, 77*, 1–14.

Inglis, M., & Alcock, L. (2012). Expert and novice approaches to reading mathematical proofs. *Journal for Research in Mathematics Education, 43*, 358–90.

Inglis, M., & Alcock, L. (2013). Skimming: A response to Weber & Mejía-Ramos. *Journal for Research in Mathematics Education, 44*, 471–4.

Inglis, M., & Mejía-Ramos, J.-P. (2009a). The effect of authority on the persuasiveness of mathematical arguments. *Cognition and Instruction, 27*, 25–50.

Inglis, M., & Mejía-Ramos, J.-P. (2009b). On the persuasiveness of visual arguments in mathematics. *Foundations of Science, 14*, 97–110.

Inglis, M., Mejía-Ramos, J.-P., Weber, K., & Alcock, L. (2013). On mathematicians' different standards when evaluating elementary proofs. *Topics in Cognitive Science, 5*, 270–82.

Inglis, M., & Simpson, A. (2008). Conditional inference and advanced mathematical study. *Educational Studies in Mathematics, 67*, 187–204.

Inglis, M., & Simpson, A. (2009). Conditional inference and advanced mathematical study: Further evidence. *Educational Studies in Mathematics, 72,* 185–98.

Johnson-Laird, P. N., & Byrne, R. M. J. (1991). *Deduction.* Hove, UK: Erlbaum.

Kember, D., & Leung, D. Y. P. (2006). Characterising a teaching and learning environment conducive to making demands on students while not making their workload excessive. *Studies in Higher Education, 29,* 165–84.

Ko, Y.-Y., & Knuth, E. (2009). Undergraduate mathematics majors' writing performance producing proofs and counterexamples about continuous functions. *Journal of Mathematical Behavior, 28,* 68–77.

Lai, Y., & Weber, K. (2014). Factors mathematicians profess to consider when presenting pedagogical proofs. *Educational Studies in Mathematics, 85,* 93–108.

Lai, Y., Weber, K., & Mejia-Ramos, J.-P. (2012). Mathematicians' perspectives on features of a good pedagogical proof. *Cognition and Instruction, 30,* 146–69.

Lakatos, I. (1976). *Proofs and refutations.* Cambridge: Cambridge University Press.

Larsen, S., & Zandieh, M. (2008). Proofs and refutations in the undergraduate mathematics classroom. *Educational Studies in Mathematics, 67,* 185–98.

Lawless, C. (2000). Using learning activities in mathematics: Workload and study time. *Studies in Higher Education, 25,* 97–111.

Leikin, R., & Wicki-Landman, G. (2000). On equivalent and non-equivalent definitions: Part 2. *For the Learning of Mathematics, 20*(2), 24–9.

Leinhardt, G., Zaslavsky, O., & Stein, M. K. (1990). Functions, graphs, and graphing: Task, learning, and teaching. *Review of Educational Research, 60,* 1–64.

Lin, F.-L., & Yang, K.-L. (2007). The reading comprehension of geometric proofs: The contribution of knowledge and reasoning. *International Journal of Science and Mathematics Education, 5,* 729–54.

Lithner, J. (2003). Students' mathematical reasoning in university textbook exercises. *Educational Studies in Mathematics, 52,* 29–55.

Lithner, J. (2008). A research framework for creative and imitative reasoning. *Educational Studies in Mathematics, 67,* 255–76.

Lizzio, A., Wilson, K., & Simons, R. (2002). University students' perceptions of the learning environment and academic outcomes: Implications for theory and practice. *Studies in Higher Education, 27*, 27–52.

Mariotti, M. A. (2006). Proof and proving in mathematics education. In A. Gutiérrez & P. Boero (Eds.), *Handbook of research on the psychology of mathematics education: Past, present and future* (pp. 173–204). Rotterdam: Sense.

Marton, F., & Säljö, R. (1976). On qualitative differences in learning 1. *British Journal of Educational Psychology, 46*, 4–11.

Mason, J., Burton, L., & Stacey, K. (1982). *Thinking mathematically*. London: Addison-Wesley.

Mason, J., & Pimm, D. (1984). Generic examples: Seeing the general in the particular. *Educational Studies in Mathematics, 15*, 277–89.

Matthews, P., & Rittle-Johnson, B. (2009). In pursuit of knowledge: Comparing self-explanations, concepts, and procedures as pedagogical tools. *Journal of Experimental Child Psychology, 104*, 1–21.

McNamara, D. S., Kintsch, E., Songer, N. B., & Kintsch, W. (1996). Are good texts always better? Interactions of text coherence, background knowledge, and levels of understanding in learning from text. *Cognition and Instruction, 14*, 1–43.

Mejía-Ramos, J.-P., Fuller, E., Weber, K., Rhoads, K., & Samkoff, A. (2012). An assessment model for proof comprehension in undergraduate mathematics. *Educational Studies in Mathematics, 79*, 3–18.

Mejía-Ramos, J.-P., & Weber, K. (2014). Why and how mathematicians read proofs: Further evidence from a survey study. *Educational Studies in Mathematics, 85*, 161–73.

Michener, E. R. (1978). Understanding understanding mathematics. *Cognitive Science, 2*, 361–83.

Mills, M. (2014). A framework for example usage in proof presentations. *Journal of Mathematical Behavior, 33*, 106–18.

Monaghan, J. (1991). Problems with the language of limits. *For the Learning of Mathematics, 11*, 20–24.

Moore, R. (1994). Making the transition to formal proof. *Educational Studies in Mathematics, 27*, 249–66.

Muis, K. R. (2004). Personal epistemology and mathematics: A critical review and synthesis of research. *Review of Educational Research, 74*, 317–77.

Nardi, E. (2008). *Amongst mathematicians: Teaching and learning mathematics at university level.* New York: Springer.

Oehrtman, M. (2009). Collapsing dimensions, physical limitation, and other student metaphors for limit concepts. *Journal for Research in Mathematics Education, 40,* 396–426.

Oehrtman, M., Swinyard, C., & Martin, J. (2014). Problems and solutions in students' reinvention of a definition for sequence convergence. *Journal of Mathematical Behavior, 33,* 131–48.

Österholm, M. (2005). Characterizing reading comprehension of mathematical texts. *Educational Studies in Mathematics, 63,* 325–46.

Peled, I., & Zaslavsky, O. (1997). Counter-examples that (only) prove and counter-examples that (also) explain. *Focus on Learning Problems in Mathematics, 19,* 49–61.

Perkin, G., Croft, T., & Lawson, D. (2013). The extent of mathematics learning support in UK higher education—the 2012 survey. *Teaching Mathematics and its Applications, 32,* 165–72.

Perry, W. G. (1970). *Forms of intellectual and ethical development in the college years: A scheme.* New York: Holt, Rinehart and Winston.

Perry, W. G. (1988). Different worlds in the same classroom. In P. Ramsden (Ed.), *Improving learning: New perspectives* (pp. 145–61). London: Kogan Page.

Pinto, M., & Tall, D. O. (2002). Building formal mathematics on visual imagery: A case study and a theory. *For the Learning of Mathematics, 22,* 2–10.

Poincaré, H. (1905). *Science and hypothesis* London: Walter Scott Publishing.

Pólya, G. (1957). *How to solve it: A new aspect of mathematical method.* Princeton, NJ: Princeton University Press.

Presmeg, N. (2006). Research on visualization in learning and teaching mathematics. In A. Gutiérrez & P. Boero (Eds.), *Handbook of research on the psychology of mathematics education: Past, present and future* (pp. 205–35). Rotterdam: Sense.

Przenioslo, M. (2005). Introducing the concept of convergence of a sequence in secondary school. *Educational Studies in Mathematics, 60,* 71–93.

Raman, M. (2003). Key ideas: What are they and how can they help us understand how people view proof? *Educational Studies in Mathematics, 52,* 319–25.

Raman, M. (2004). Epistemological messages conveyed by three high-school and college mathematics textbooks. *Journal of Mathematical Behavior, 23,* 389–404.

Rav, Y. (1999). Why do we prove theorems? *Philosophia Mathematica, 7,* 5–41.

Recio, A., & Godino, J. (2001). Institutional and personal meanings of mathematical proof. *Educational Studies in Mathematics, 48,* 83–99.

Renkl, A. (2002). Worked-out examples: Instructional explanations support learning by self-explanations. *Learning and Instruction, 12,* 529–56.

Rittle-Johnson, B. (2006). Promoting transfer: Effects of self-explanation and direct instruction. *Child Development, 77,* 1–15.

Robert, A., & Speer, N. (2001). Research on the teaching and learning of calculus/elementary analysis. In D. Holton (Ed.), *The teaching and learning of mathematics at university level* (pp. 283–99). New York: Springer.

Roh, K. H. (2008). Students' images and their understanding of definitions of the limit of a sequence. *Educational Studies in Mathematics, 69,* 217–33.

Rowland, T. (2002). Generic proofs in number theory. In S. R. Campbell & R. Zazkis (Eds.), *Learning and teaching number theory: Research in cognition and instruction* (pp. 157–84). Westport, CT: Ablex Publishing Corp.

Roy, M., & Chi, M. T. H. (2005). The self-explanation principle in multimedia learning. In E. Mayer (Ed.), *The Cambridge handbook of multimedia learning* (pp. 271–86). Cambridge: Cambridge University Press.

Schoenfeld, A. H. (1985). *Mathematical problem solving.* San Diego: Academic Press.

Schoenfeld, A. H. (1992). Learning to think mathematically: Problem solving, metacognition and sense making in mathematics. In D. Grouws (Ed.), *Handbook of research on mathematics teaching and learning* (pp. 334–70). New York: Macmillan.

Sealey, V. (2014). A framework for characterizing student understanding of Riemann sums and definite integrals. *Journal of Mathematical Behavior, 33,* 230–45.

Segal, J. (2000). Learning about mathematical proof: Conviction and validity. *Journal of Mathematical Behavior, 18,* 191–210.

Selden, A., & Selden, J. (1999). *The role of logic in the validation of mathematical proofs* (Tech. Rep.). Cookeville, TN, USA: Tennessee Technological University.

Selden, A., & Selden, J. (2003). Validations of proofs considered as texts: Can undergraduates tell whether an argument proves a theorem? *Journal for Research in Mathematics Education, 34*, 4–36.

Selden, J., & Selden, A. (1995). Unpacking the logic of mathematical statements. *Educational Studies in Mathematics, 29*, 123–51.

Shepherd, M. D. (2005). Encouraging students to read mathematics. *Problems, Resources, and Issues in Mathematics Undergraduate Studies, 15*, 124–44.

Shepherd, M. D., Selden, A., & Selden, J. (2012). University students' reading of their first-year mathematics textbooks. *Mathematical Thinking and Learning, 14*, 226–56.

Skemp, R. R. (1976). Relational understanding and instrumental understanding. *Mathematics Teaching, 77*, 20–26.

Sofronas, K. S., DeFranco, T. C., Vinsonhaler, C., Gorgievski, N., Schroeder, L., & Hamelin, C. (2011). What does it mean for a student to understand the first-year calculus? Perspectives of 24 experts. *Journal of Mathematical Behavior, 30*, 131–48.

Speer, N. M., Smith III, J. P., & Horvath, A. (2010). Collegiate mathematics teaching: An unexamined practice. *Journal of Mathematical Behavior, 29*, 99–114.

Stanovich, K. E. (1999). *Who is rational? Studies of individual differences in reasoning.* Mahwah, NJ: Lawrence Erlbaum.

Stewart, I. N., & Tall, D. O. (1977). *The foundations of mathematics.* Oxford: Oxford University Press.

Stylianides, A. J., & Stylianides, G. J. (2009). Proof constructions and evaluations. *Educational Studies in Mathematics, 72*, 237–53.

Stylianides, A. J., Stylianides, G. J., & Philippou, G. N. (2004). Undergraduate students' understanding of the contraposition equivalence rule in symbolic and verbal contexts. *Educational Studies in Mathematics, 55*, 133–62.

Stylianou, D. A., & Silver, E. A. (2004). The role of visual representations in advanced mathematical problem solving: An examination of expert–novice similarities and differences. *Mathematical Thinking and Learning, 6*, 353–87.

Swinyard, C. (2011). Reinventing the formal definition of limit: The case of Amy and Mike. *Journal of Mathematical Behavior, 30*, 93–114.

Tall, D. (1982). Elementary axioms and pictures for infinitesimal calculus. *Bulletin of the IMA, 18*, 43–83.

Tall, D. (1991). Intuition and rigour: The role of visualization in the calculus. In W. Zimmerman & S. Cunningham (Eds.), *Visualization in teaching and learning mathematics* (pp. 105–19). Washington, DC: MAA.

Tall, D. (2013). *How humans learn to think mathematically.* Cambridge: Cambridge University Press.

Tall, D. O. (1989). The nature of mathematical proof. *Mathematics Teaching, 127*, 28–32.

Tall, D. O. (1992). The transition to advanced mathematical thinking: Functions, limits, infinity, and proof. In D. A. Grouws (Ed.), *Handbook of research on mathematics teaching and learning* (pp. 495–511). New York: Macmillan.

Tall, D. O. (1995). Cognitive development, representations and proof. In *Proceedings of justifying and proving in school mathematics* (pp. 27–38). London: Institute of Education.

Tall, D. O., & Vinner, S. (1981). Concept image and concept definition in mathematics with particular reference to limits and continuity. *Educational Studies in Mathematics, 12*, 151–69.

Toulmin, S. (1958). *The uses of argument.* Cambridge: Cambridge University Press.

Tsamir, P., Tirosh, D., & Levenson, E. (2008). Intuitive nonexamples: The case of triangles. *Educational Studies in Mathematics, 49*, 81–95.

Vamvakoussi, X., Christou, K. P., Mertens, L., & Van Dooren, W. (2011). What fills the gap between discrete and dense? Greek and Flemish students' understanding of density. *Learning and Instruction, 21*, 676–85.

Vamvakoussi, X., & Vosniadou, S. (2010). How many decimals are there between two fractions? Aspects of secondary school students' understanding of rational numbers and their notation. *Cognition and Instruction, 28*, 181–209.

Vinner, S. (1991). The role of definitions in teaching and learning. In D. O. Tall (Ed.), *Advanced mathematical thinking* (pp. 65–81). Dordrecht: Kluwer.

Vinner, S., & Dreyfus, T. (1989). Images and definitions for the concept of function. *Journal for Research in Mathematics Education, 20*, 356–66.

Weber, K. (2001). Student difficulty in constructing proofs: The need for strategic knowledge. *Educational Studies in Mathematics, 48*, 101–19.

Weber, K. (2004). Traditional instruction in advanced mathematics courses: A case study of one professor's lectures and proofs in an introductory real analysis course. *Journal of Mathematical Behavior, 23*, 115–33.

Weber, K. (2005). On logical thinking in mathematics classrooms. *For the Learning of Mathematics, 25*(3), 30–31.

Weber, K. (2008). How mathematicians determine if an argument is a valid proof. *Journal for Research in Mathematics Education, 39*, 431–59.

Weber, K. (2009). How syntactic reasoners can develop understanding, evaluate conjectures, and generate examples in advanced mathematics. *Journal of Mathematical Behavior, 28*, 200–208.

Weber, K. (2010a). Mathematics majors' perceptions of conviction, validity and proof. *Mathematical Thinking and Learning, 12*, 306–36.

Weber, K. (2010b). Proofs that develop insight. *For the Learning of Mathematics, 30*(1), 32–6.

Weber, K. (2012). Mathematicians' perspectives on their pedagogical practices with respect to proof. *International Journal of Mathematical Education in Science and Technology, 43*, 463–82.

Weber, K., & Alcock, L. (2004). Semantic and syntactic proof productions. *Educational Studies in Mathematics, 56*, 209–34.

Weber, K., & Alcock, L. (2005). Using warranted implications to understand and validate proofs. *For the Learning of Mathematics, 25*(1), 34–8.

Weber, K., & Alcock, L. (2009). Proof in advanced mathematics classes: Semantic and syntactic reasoning in the representation system of proof. In D. A. Stylianou, M. L. Blanton, & E. Knuth (Eds.), *Teaching and learning proof across the grades: A K-16 perspective* (pp. 323–38). New York: Routledge.

Weber, K., Inglis, M., & Mejía-Ramos, J.-P. (2014). How mathematicians obtain conviction: Implications for mathematics instruction and research on epistemic cognition. *Educational Psychologist, 49*, 36–58.

Weber, K., & Mejía-Ramos, J.-P (2009). An alternative framework to evaluate proof productions: A reply to Alcock and Inglis. *Journal of Mathematical Behavior, 28*, 212–16.

Weber, K., & Mejía-Ramos, J.-P (2011). Why and how mathematicians read proofs: An exploratory study. *Educational Studies in Mathematics, 76*, 329–44.

Weinberg, A., & Wiesner, E. (2011). Understanding mathematics textbooks through reader-oriented theory. *Educational Studies in Mathematics, 76*, 49–63.

Weinberg, A., Wiesner, E., & Fukawa-Connelly, T. (2014). Students' sense-making frames in mathematics lectures. *Journal of Mathematical Behavior, 33*, 168–79.

Wicki-Landman, G., & Leikin, R. (2000). On equivalent and non-equivalent definitions: Part 1. *For the Learning of Mathematics, 20*(1), 17–21.

Williams, C. G. (1998). Using concept maps to assess conceptual knowledge of function. *Journal for Research in Mathematics Education, 29*, 414–21.

Wong, R. M. F., Lawson, M. J., & Keeves, J. (2002). The effects of self-explanation training on students' problem solving in high-school mathematics. *Learning and Instruction, 12*, 233–62.

Yang, K.-L., & Lin, F.-L. (2008). A model of reading comprehension of geometry proof. *Educational Studies in Mathematics, 67*, 59–76.

Yeager, D. S., & Dweck, C. S. (2012). Mindsets that promote resilience: When students believe that personal characteristics can be developed. *Educational Psychologist, 47*, 302–14.

Yopp, D. A. (2014). Undergraduates' use of examples in online discussion. *Journal of Mathematical Behavior, 33*, 180–91.

Yusof, Y. B. M., & Tall, D. O. (1999). Changing attitudes to university mathematics through problem solving. *Educational Studies in Mathematics, 37*, 67–82.

Zandieh, M., & Rasmussen, C. (2010). Defining as a mathematical activity: A framework for characterizing progress from informal to more formal ways of reasoning. *Journal of Mathematical Behavior, 29*, 57–75.

Zandieh, M., Roh, K. H., & Knapp, J. (2014). Conceptual blending: Student reasoning when proving 'conditional implies conditional' statements. *Journal of Mathematical Behavior, 33*, 209–29.

Zaslavsky, O., & Shir, K. (2005). Students' conceptions of a mathematical definition. *Journal for Research in Mathematics Education, 36*, 317–46.

Zazkis, R., & Chernoff, E. J. (2008). What makes a counterexample exemplary? *Educational Studies in Mathematics, 68*, 195–208.

INDEX